Communication Skills for Engineers

Second Edition

Dr Sunita Mishra
University of Hyderabad

Dr C. Muralikrishna
Osmania University

Associate Acquisitions Editor: Sandhya Jayadev
Associate Production Editor: M. R. Ramesh
Composition: Sigma Publishing Services, Chennai
Printer:

ISBN 978-81-317-3384-4

10 9 8 7 6 5 4 3 2

Published by Pearson India Education Services Pvt. Ltd, CIN: U72200TN2005PTC057128

Head Office: 1st Floor, Berger Tower, Plot No. C-001A/2, Sector 16B, Noida – 201 301,Uttar Pradesh, India.

Registered Office: Featherlite, 'The Address' 5th Floor, Survey No 203/10B, 200 Ft MMRD Road, Zamin Pallavaram, Chennai - 600044
Website: in.pearson.com; Email:companysecretary.india@pearson.com

Digitally printed in India by Trinity Academy for Corporate Training Ltd, New Delhi in the year of 2025.

CONTENTS

PREFACE

The learning and teaching of communications has generally been limited to the spoken and the written skills—presentations, group discussions, writing of letters, reports, etc. For students enrolled in professional courses, it extends to document writing or maybe, project writing. Experience, however, shows that a good communicator has more than just these skill sets. It is primarily an attitude, a willingness to communicate, share one's ideas and information that makes one a good communicator. Language and the knowledge of the various formalities associated with speaking and writing do matter. However, given the right attitudinal input, communication becomes much easier and one emerges as an effective communicator.

This book discusses additionally these attitudinal factors that make one a good communicator and links them up with the skill sets that enable effective communication. It speaks of the writing and the speaking skills in the context of creativity, negotiation, interpersonal skills and problem solving. Specifically, the book aims at developing the communication skills of Engineering and other professional students. For this target group, good communication skills are necessary for recruitment and to enhance their opportunity for further growth in the profession.

The book attempts to locate the common communication needs of this group in different situations and to guide and equip the readers and learners to fulfill them appropriately. The book follows a skill-based approach. It isolates the skill sets required in different communicative situations, gives the students a comprehensive view of the requirements and finally reinforces them further through exercises and activities.

Communication Skills for Engineers (CSE) thus is a comprehensive book that focuses on the communication needs of users from the Engineering and other professional areas. It does not look at "communications"—fluency in speaking and writing—in isolation but discusses the whole attitudinal framework that enables effective and purposeful exchange of information. It aims at enabling students to identify and develop skill sets necessary to succeed in their profession.

This second edition of CSE covers some additional topics that include a discussion of grammar dynamics and activities thereof; information technology (IT) and communication; the nitty-gritty of e-mail writing, interview skills; designing SOPs, a CV and Resumé writing, and aspects of improving reading skills among others. This has been done to cater to a wider range of felt needs among Engineering students and other professional students. To accommodate these additions, some of the earlier chapters have been collated appropriately. In this process, we have tried our best to keep the book's main points as well as its pedagogy intact.

We hope that the second edition of CSE will continue to help the students, teachers and trainers from Engineering and other professional streams.

Acknowledgements

This book is a combined effort of many minds. It was conceived when we were teaching MBA, and MCA students and took shape in the course of our lectures to B.Tech., B.E., B.A., B.Sc. and B.Com. graduates. However, many influences have gone into its making. We would specially like to thank the following people:

- Our teacher, Prof. N. Krishnaswamy, for his encouraging influence and useful advice. He was instrumental in planting the idea of this book in our minds.

- Late Prof. G. V. Raj for his valuable role in giving us the English language teaching (ELT) orientation in our early days of teaching.
- Our students from Osmania University, Jawaharlal Nehru Technical University (JNTU) and University of Hyderabad, whose responses and reactions helped to shape the contents of the book in a big way.
- Prof. Tutun Mukherjee for her timely appreciation and for her assessment of our book proposal.
- Dear friends like Vijaya, Joy, Seetha and Srinivas for goading us to persevere in our efforts and giving valuable suggestions and ideas.
- K. Srinivas, the then commissioning editor of Pearson Education who is now the publishing manager, for his relevant, practical and timely suggestions in the formative stages of the book.
- Brig. Sambhi, former Principal of Karshak Engineering College, for his creative ideas and suggestions.
- Our dear daughter, Shivani, for her growing understanding, and support and for having put up with our preoccupied minds during the long months of writing.

Dr Sunita Mishra
Prof. C. Muralikrishna

INTRODUCTION

Information Technology and Communication

Objectives

This introductory section takes an overview of the role of Information Technology in communication and also on the subject of communication. It discusses the different aspects and issues relating to communication. It underpins the value of creativity in communication in addition to delving on interpersonal communication, body language, persuasive communication and negotiation.

INFORMATION TECHNOLOGY IN INDIA AND THE PROCESS OF COMMUNICATION

Many outstanding technologies have come into the lives of people towards the end of the 20th century. This is especially true in the field of electronics. Many electronic types of equipment like mainframe computers, mini computers, personal computers, e-mail, cell phones, i-pods, and other gadgets have become part of our lives. In this informaton age, we come across a lot of personal or office data that needs to be processed by computers and this is called information technology or IT.

Information technology (IT) accounts for a significant part of India's GDP and export earnings, while providing employment to a large number of its tertiary sector workforce. Technically equipped immigrants from India have been seeking jobs in the western world since the 1950s as India's education system produced more engineers than its industry could absorb. Thus, India's growing stature in the information age enabled it to form close relations with both the United States of America and the European Union.

Out of around 400,000 engineers produced per year in the country, approximately, about 100,000 are well equipped in both technical competency and the English language skills. India has developed a number of outsourcing companies specializing in customer support through the Internet or telephone. For instance, by 2008, India had a total of 49,750,000 telephone lines in use, a total of 233,620,000 mobile phone connections, a total of 60,000,000 Internet users—comprising 6.0% of the country's population, and 4,010,000 people in the country had access to broadband Internet— making it the 18th largest country in the world in terms of broadband Internet users. Total fixed-line and wireless subscribers reached 325.78 million in June, 2008.

On the other hand, from the dynamics point of view, it needs to be mentioned that Management Information Systems (MIS) has assumed vital importance. MIS is the application of IT to the communication process in organizations. Generally, in MIS most of the information processing is done by the computers. The span of activities undertaken by MIS includes among others generating, processing and transmitting information. MIS is used for interpersonal and organizational communication in addition to strategy planning and improving customer care. Thus, managers can seek vital clarifications, information from and give instructions and directions to fellow-personnel of the organization through the systems.

An important effect of IT on organizations is one of leveling. Personnel in organizations, whether they are superiors or subordinates, work in an information-sharing ambience where everybody is involved non-hierarchically in the developmental work of the organization. Interestingly, the IT age continues to demand proficiency in English communication skills from most of its initiators, mid-level force and end-users. It is in this context that one needs to look at the nitty-gritty of communication.

COMMUNICATION: ASPECTS AND ISSUES

> When American playwright and diplomat Clare Boothe Luce visited George Bernard Shaw in his London flat years ago, she found him writing at his desk. To express her admiration, she began, "Mr Shaw, you are the only reason I'm standing here."
>
> Shaw looked up and replied, "Who did you say your mother was, my child?"
>
> (Reader's Digest)

Communication, as we see it here, can be a complicated process of give-and-take with innumerable intricacies and dimensions. Often, however, it is seen as a set of competencies, primarily including the written and the spoken mode. It is taught as a package involving training in skills pertaining to writing, speaking, reading and listening (L. S. R. W). But general observation shows that effective communication involves a lot more than proficiency in the L. S. R. W skills. More than language, it needs an attitude, a willingness to give and take; to open up to others and accept others; to have empathy and capacity to look at situations from varied perspectives. Given these attitudinal factors, language becomes just an aid to promote communication. This chapter aims to give an overall view of most of these important factors and show how they enable communication.

A complicated process in itself, communication takes place all around us all the time. In fact, we all spend around 70% of our time receiving or sending messages. Essentially, it involves the sender or the communicator, and the receiver. It is sent in a certain medium through encoded messages. The receiver, in turn, decodes the message and sends back the reactions to the sender in the form of feedback. The beauty of the whole process, however, lies in the nature of communication itself. Language, be it in any form, has the potential to mean many things at the same time. Thus, the sender sends a message, which the receiver is free to make meaning of depending on the mode of transmission, the kind of encoding and of course, more importantly, the receiver's own state of mind while receiving the message.

VITALS OF COMMUNICATION

The features that have been looked into in this section are, function of language and elements of human communication.

The Function of Language

Language embodies and conveys thought. It is an important means that we rely on to convey our thoughts and feelings. In its spoken and written forms, language is the commonest and most important means of communication in all social activities among human beings. Along with language there are other elements, which contribute to communication. In the following section, some of these elements are briefly examined.

Elements of Human Communication

Human communication, as a process, is not a dull and passive process. It is a dynamic and active process comprising several important elements. These elements may be enumerated as follows:

1. *Initiation*: Communication begins when a source initiates a statement. A statement is initiated in order to transmit some thought, need, idea or information. The receiver attends to the statement transmitted by the source, interprets the statement and decides how to respond.
2. *Feedback*: The response of the receiver that is sent back to the source forms the feedback. The source modifies further statements based on the feedback. Feedback helps the source to know if the message was received correctly or not.
3. *Channel*: Channel connects the source (e.g. a speaker) and the receiver (e.g. listener). A speaker and a listener are connected to each other by sound waves and (or) the light waves. That is, language carried by sound waves; and facial expressions and body gestures carried by light waves.
4. *Situation*: Situation is the place or setting in which a communicative event occurs.
5. *Purpose*: Purpose consists of the intention of the source, or speaker. It is the communicative aim of the speaker.
6. *Attitudes*: The speaker and the listener, carry with them certain ideologies, world-views, beliefs, likes, dislikes and aptitudes. They are also under the influence of varying emotional and mental states. These factors affect the attitudes of the speaker and the listener at the time of communication.
7. *Knowledge*: The speaker has to possess adequate knowledge of the message that is to be transmitted. Knowledge that is based on observation, study and personal experience helps the speaker to communicate effectively.
8. *Expression*: Expression consists of the ability to transmit or communicate. Fluency, clarity and intelligibility of expression pave the way to effective communication. The flow of information, as embodied in the message that is transmitted, is smooth when expression

is clear. This helps the speaker and the listener to avoid communication gaps and also arrive at consensus and decisions. Improper or faulty expression leads to breakdown of communication.

9. *Language*: Language is one of the most important elements of the communication process. The effective use of language consists in selecting appropriate words and patterns of sentences while communicating. These linguistic patterns, suitably supported by facial and body gestures, enable effective communication.
10. *Intellectualism*: Communication is sustained and it becomes effective only in an intellectual ambience. That is, the speaker and the listener have to express and understand views calmly, rationally, reflectively, precisely and efficiently. When intellectualism is absent, thoughts and ideas are likely to be ineffectual.

See What This Communicates!

Once upon a time, there were four people. Their names were Everybody, Somebody, Nobody and Anybody. Whenever there was an important job to be done, Everybody was sure that Somebody would do it. Anybody could have done it, but Nobody did it. Everybody got angry because it was Everybody's job. Everybody thought that Somebody would do it, but Nobody realized that nobody would do it. So consequently, Everybody blamed Somebody when nobody did what Anybody could have done in the first place.

ACTIVITIES

1. Read the sentences below (A) and see in how many different contexts you can use them. Create small contexts around them and use them to illustrate the given modes of expression(B). Add more to the list if you can.

A	B
a. The door is open.	Giving information Making an offer Rejection Acceptance
b. You are joking! This can't be true.	Extreme happiness Shock Anger
c. The possibilities are endless.	Appreciation /Frustration Amazement /Confusion

2. Try to read the following groups of sentences separately stressing the underlined word. Make a note of the different meanings that emerge.

(A)
Even he does not write regularly.
Even he does not write regularly.
Even he does not write regularly.

(B)
You don't know me?
You don't know me?
You don't know me?

(C)
Yes. I don't remember.
Yes. I don't remember
Yes. I don't remember

(D)
Is this possible?
Is this possible?

(E)
I can't write the examination.
I can't write the examination
I can't write the examination

CREATIVITY IN COMMUNICATION

While doing the above exercises, you must have realized that responding to a communicational situation needs creativity. The meaning has to be always put together like the pieces of a jigsaw puzzle. One always has to "make" the final picture for oneself. When we talk of creativity in communication, we mean the capacity to read hidden messages, to see a link not seen before, to infer the unsaid, unspelt consequences or even to tailor a response that can alter a communication situation to one's advantage! A witty creative receiver can create meanings in the encoded message the sender had never imagined. Creativity in day-to-day interactions is also one's ability to break out of stereotyped reactions, to see beyond the set, accepted patterns and give a novel twist to existing "reality" or better still, create a "reality"!

Albert Einstein was travelling to universities in a chauffeur driven car, delivering lectures on his theory of relativity. One day while in transit, the chauffeur remarked:

"Dr.Einstein, I have heard you deliver that lecture thirty times. I know it by heart and bet I could give it myself". "Well, I will give you the chance," said Einstein. "They don't know me at the next college, so when we get there I'll put on your cap, and you introduce yourself as me and give the lecture". The chauffeur delivered Einstein's lecture flawlessly. When he finished, he started to leave, but one of the professors stopped him and asked a complex question filled with mathematical equations and formulas. The chauffeur thought fast. "The solution to that problem is so simple", he said, "I'm surprised you have to ask me. In fact, to show you just how simple it is, I'm going to ask my chauffeur to come up here and answer your question."

—*Readers Digest*

ACTIVITIES

3. Given below, is a list of ten sentences. Take one minute to think about each and talk on it for a couple of minutes.

   The sky is blue
   The height of success.
   The ceiling is very high.
   Bicycling under water is fun.
   Classes are not to be attended.
   It is fun to fail.
   Among the stars.
   Children are mean.
   The blackboard is green.
   My useless fridge.

4. Try to establish links between the pairs of objects given and talk about each of the pairs.

   Apple // Computer
   Horse // House
   Armchair // Fridge
   Star // College
   Pigs // Saints
   Trees // Chocolates
   Hill // Eagle
   Insect // Money
   Bird // Ceiling
   Chair // Piano.

COMMUNICATING WITH CONCERN AND EMPATHY

Very often we come across managers who would say: "I do a lot for my employees. I expect a lot out of them. I work hard to be friendly towards them and treat them right. However, I feel they are ungrateful. I think if I stayed back at home for a single day they would waste time at the office canteen. What can I do to make them independent and responsible?"

Problems like this are common. A large part of it is because we fail to communicate our concern or empathy. A very closely related quality that plays an equally important role in communication is **empathy**. **Empathy** is a quality that will allow you to understand the mental and psychological state of the person we are communicating with. This will help us to adapt our response and pitch it at an acceptable and appropriate level. Important here is the concept of **empathetic listening**, which can be explained as listening with a view to understand. We generally listen with an intent to reply. Most of the times, we are either talking or getting ready to talk. When we listen, we generally listen at one of the four levels. We may be **ignoring** the person, not really listening at all. We may be **pretending** to listen. We may be doing **selective listening**, hearing only certain parts of the conversation, or we may be practising **attentive listening**, paying attention to every word that is being uttered with utmost care. What is required often, however, is **empathetic listening** that forms the fifth grade of listening. This is something beyond the skill-based acquisition of the skill. Here listening is done with a purpose to understand, to see the world from the speaker's perspective intellectually and emotionally, but not necessarily agree with the speaker. This kind of listening has an almost therapeutic effect on both the speaker and the listener, primarily because it gives a person a "psychological air". This "psychological air" deeply impacts communication, expanding both the speaker's and the listener's area of influence on one another. Paradoxically, this kind of listening is also risky. To listen deeply, one also has to open oneself up and become vulnerable. To have influence, thus, one has to risk being influenced.

> The poet doesn't invent. He listens.
>
> —*Jean Cocteau*

Listening to a Disturbed Person

At some point or the other, all of us have been disturbed or we have interacted with people who are disturbed. It could be a friend who feels that injustice has been done to him/her at the workplace, or a neighbour whose feelings have been wounded.

Some people are good at expressing their hurt feelings, some are not. Sometimes, we even try to suppress them and push them to the back of our mind. But unexpressed feelings do not disappear. They are like springs. You push them down but they express themselves sooner or later, may be at the wrong place or time.

A person who has suppressed feelings cannot act or think straight because he/she does not feel straight. Only after the feelings are released can the person think straight and look into the problem objectively. Above everything else, more than advice, evaluation or judgement, a disturbed friend needs to fulfill the desire to be understood.

Do not listen to their words, listen to their feelings. Respond with the whole of yourself. The best way to 'talk' to a disturbed person is to listen.

THE JOHARI WINDOW

The concept of the **Johari window** is an important idea that explains factors behind mutual understanding. Named after **Joseph Luft** and **Harry Ingram**, it explains communications along two dimensions: exposure and feedback. Exposure is the extent to which the individual is willing to divulge his feelings and information in trying to communicate. Feedback is the extent to which the individual manages to elicit exposure from others. They translate these factors into four windows—open, blind, hidden and unknown (see Fig.1). The "open" window is the information about you, which you as well as others can see. The "blind" window encompasses the factors about you that others can see but not you. This, often, is the result of your defensive behaviour that prevents others from telling you things about yourself. The "hidden" window is the information known to you but unknown to others. These include facts about yourself that you hide from others. Finally, the "unknown" constitutes factors that that neither you nor others are aware of. Advocates of the Johari window see openness, authenticity and honesty as valued qualities in interpersonal relationship. They also imply that it is in the interest of the individual to expand the size of the "open" window, increase self-disclosure and be more willing to listen to feedback from others about oneself.

Open	Blind
Hidden	Unknown

Figure 1: Johari Window

ACTIVITIES

5. Select a relationship in your life where your emotional account is nil. Write five reasons from your perspective on why you think the situation is so. Now try and approach the problem from the other person's point of view and review the situation. Were your assumptions all correct?

6. Give five reasons for and against the statements given below.

1. Siblings can be your best friends because,
   a. ____________
   b. ____________
   c. ____________
   d. ____________
   e ____________

   Siblings are one's greatest enemies because,
   a. ____________
   b. ____________
   c. ____________
   d. ____________
   e. ____________

2. You are your greatest friend because,
   a. ____________
   b. ____________
   c. ____________
   d. ____________
   e. ____________

   You are your greatest enemy because,
   a. ____________
   b. ____________
   c. ____________
   d. ____________
   e. ____________

3. Holidays are wonderful
   a. ________________
   b. ________________
   c. ________________
   d. ________________
   e. ________________

   Holidays are not wonderful
   a. ________________
   b. ________________
   c. ________________
   d. ________________
   e. ________________

4. Higher education should be subsidized
   a. ________________
   b. ________________
   c. ________________
   d. ________________
   e. ________________

   Higher education should not be subsidized
   a. ________________
   b. ________________
   c. ________________
   d. ________________
   e. ________________

5. Children are cruel
   a. ________________
   b. ________________
   c. ________________
   d. ________________
   e ________________

   Children are not cruel
   a. ________________
   b. ________________
   c. ________________
   d. ________________
   e. ________________

7. Recall any incident in your life, which brought forth in you something that, belonged to the "unknown window" till then. How have you been managing that aspect/quality since then?

8. Imagine, you have just died and your friends want to put up your picture in your class with the following words written below:

   "Here was a person who… "
   Complete the paragraph in about 30 words.

INTERPERSONAL COMMUNICATION

> *I can live two months on a good compliment.*
>
> *—Mark Twain*

Possessing good knowledge of interpersonal relations and being able to achieve the delicate balance that communication requires goes a long way in fine-tuning our reflexes to achieve effective communication. For instance, it would be extremely inappropriate on our part to complain to a new recruit at our office against the boss or a nagging spouse at home. It is very important to be able to decide who should be told what, when and how.

Communication is generally divided into five levels, depending on the nature, scope and depth of interaction:

Passing Communication

This refers to the daily niceties or phatic communication patterns like "hello, good morning" and "good night" etc. The purpose here is to merely acknowledge the person's presence. Here a detailed response is neither expected nor given. When somebody wishes you in office saying "Hello! How are you?" you definitely do not sit back and say, "Not well at all! Know what happened on the road today?..."

Factual Communication

Often, most of the organizational work depends on the exchange of factual information. Giving instructions, evaluating and passing on information accurately and concisely fall under this category. Interpersonal communication is carried on here with minimal possible emotional risk. Examples of such communication are: Is the report ready? Has the new employee joined today? How many more customers are there to meet?

Thoughts and Ideas

Here, the exchange is at a slightly higher level of communication. This includes exchange of ideas that suggests a more involved interpersonal communication. However, it also throws open the possibility of rejections – hence, the risk involved is more.

Feelings

The feeling level of communication involves a higher degree of risk. During this kind of interaction, there is exchange of sentiments and feelings making the sender vulnerable to rejection.

Peak Communication

This is the highest level of communication that is most difficult to achieve. It ensures perfect understanding between two individuals or a group of individuals. Creativity and a synchronized work culture are the building blocks of this kind of communication.

> *A bore is a man who, when*
> *you ask him how he is, tells you.*
>
> *—Bert Leston*

Activities

9. Look at the statements given below and categorize them into the different communication levels.
   1. Oh God! It's late already.

2. Will you ever learn to see straight?
3. I think we have reached the end of the road.
4. How do you do?
5. It's rather late, isn't it?
6. I agree that capitalism has its merits but we can't overlook its demerits.
7. We all agree. We can achieve this together.
8. Have you received any answer to the letter?
9. It is a pleasure working together. How else do you think we have survived our boss?
`10. I strongly believe that the developed countries are causing more harm to humanity than the underdeveloped ones.

10. Complete the following incomplete dialogue using three different situations illustrating the 2nd and 3rd levels of communication:

A. ____________________

B. Oh yes! I know.

A. ____________________

B. ____________________

A . Sure. I'll remember.

People who feel good about themselves produce good results.
—Henneth Blanchard and Spencer Johnson.

BODY LANGUAGE

Body language is an important factor that one has to keep in mind while communicating. Most of our communication in every day matters takes place through body language. Most important are the gestures, the tone and the facial expression. If the spoken words do not match the body language, it is the message covered by the body language that creates the final and lasting impression.

The body language of the person who is being spoken to is of immense importance for a good communicator. It is an important feedback not only to decide how one's message has been accepted but also to determine whether it is the right time to convey the message at all!

Arms folded around one's chest, for example, can signify a defensive attitude. It could mean "I don't want to know" or "I feel vulnerable as you talk to me". If the fists are clenched, in addition, it suggests holding back of emotion or information; finger on the lips while speaking can show incongruence of thought or speech. Looking away while someone is speaking or rubbing the eye definitely suggests avoidance. The sitting posture also can signify the attitude of the speaker. Reversing the chair and sitting, or leaning over the back can indicate power and control. Slumping with arms folded or clasped on the lap can convey dejection or submissiveness, while leaning back on the chair with legs crossed and hands behind the back signify authority, the "I don't need to fear you at all!" attitude.

While it is necessary to consider body language as an important clue in the communicative process, one should also remember that it is largely situational. They do not mean anything by themselves, but acquire their precise meaning only in association with other symbols.

Activity

11. Given below is a list of ten activities. Form groups of two and enact the situations without talking. Ask the class to guess the situation. Discuss the responses and clues on the basis of which they formed their opinion.

    a. A father is indignant because he feels his son is going astray and not concentrating on his studies. The son feels his father is unnecessarily pressurizing him.
    b. Two friends have met after a long time at a party. They are overjoyed to find one another and are engrossed in their talk.
    c. You are trying your best to convince your client that the software you have newly developed will work. He, however, is sceptical about the whole product.
    d. Your teacher is trying hard to explain a concept to you. In between, he forgets a name and is trying hard to recollect it.
    e. Your subordinate is trying hard to flatter and praise you. You, however, are very suspicious of his motive .You want to tell him so, but you are holding back out of courtesy.
    f. Your friend has come to you with a grave professional problem. You are thinking hard of the possible solutions. Your friend, too, is troubled.
    g. You are a junior employee in a company. The company is facing a financial crisis and laying off people. You have been called into the boss's room and asked to sit. The boss looks grave and you are nervous.
    h. A colleague is very enthusiastically trying to explain a plan to you. You are alert, eager and taking in every word.
    i. You are wild with your employee but trying to control your anger. Your employee is conscious of his guilt and is apprehensive.
    j. You are worried, tensed and very confused. Your friend is trying to cool you down and comfort you. She is strong, cool and confident, but very concerned.

PERSUASIVE COMMUNICATION AND NEGOTIATION

> In the film titled "The Shop Around the Corner", a character says, "when my boss calls me an idiot, I agree. After all, I'm no fool."
>
> —*Readers Digest.*

Often we judge situations, decide our role in the context and react in the way we consider appropriate. A very important skill that is used subtly in all communication situations is "negotiation". To be able to negotiate during communications, one has to be first conscious of one's role and also see how others perceive it. This is followed by each participant demarcating his area of operation and negotiating space, and deciding whether and how much to empower the other participant. The next stage is one where the exchange is decided. The terms and conditions here are largely a fallout of the first two factors.

The nature of exchange one proposes or accepts depends on the role one has fixed for oneself and also the degree of empowerment that has been negotiated and agreed upon. The next and final stage can be called the disclosure stage that comes by if the first three stages have been positively set. At this stage, participants let down their barriers and are willing to disclose information about themselves. This stage is risky, but it also paves way for successful "peak" communication. Participants moving through exchange and disclosure stage communicate with a high sense of interpersonal rapport.

ROLES WE TAKE ON DURING NEGOTIATION

Psychologists believe that the roles we take up during interaction are either that of the parent, the adult or the child. The *parent* role largely constitutes the "that is how it is" attitude. It also includes admonitions, giving rules, laws and value judgments. The *child* is the impulsive and emotional aspect of a person that can throw tantrums and get depressed. Simultaneously, it also includes creativity and curiosity of a person. The third side to an individual is his *adult* – the aspect that weighs, decides and displays appropriate emotions and expressions. During interpersonal communication, transaction can take place between similar roles in individuals or between different roles. The negotiation could be between the parents in two individuals, between the parent in one and child in another or between the adult in one and parent in the other. What is important here is that, during negotiations one should be conscious of the role he himself is playing, recognize the role the other is adopting and react accordingly. Some of the factors that play an important role while negotiating successfully are given below:

- Be conscious of the way you and your participants have positioned yourselves and the factors that each is looking for in the communication situation.
- It helps often if the participants are fair and willing to commit themselves. A communication situation built on a sense of fairness and trust can result in strong, satisfying, lasting relationships.
- We often conclude that our win depends on somebody's loss and we work towards a win-loose situation. Effective communication situations however are often built around win-win situations. The situation allows both parties to benefit or empower themselves, either materially or emotionally.

Two Most Common Traits of Managers Who Failed

1. Rigidity: they were unable to adapt their style to changes. They were unable to respond to feedback. They couldn't listen or learn.
2. Poor relationships: they were harshly critical, insensitive or demanding. They alienated those when they worked with.

Activity

12. Given below are two situations. Analyze them, discuss what kind of negotiation situation they project and think of solutions.
    a. An old piano is for sale. The prospective buyer and the seller agree on the cost. The buyer first wants the chair and then the musical pages to be given free. The seller agrees. Encouraged, the buyer next wants the delivery also to be done free.
    b. Ravi is the head of a designing team in a company producing electrical equipments. He has a good team, he gives innovative ideas and leads them from the center. Of late, however, he has been experiencing burn out, and exhaustion. He has spoken to the boss and arranged for a two-week vacation with his family. A week before leaving, the company gets the news that top executives from an internationally reputed company will be coming for a visit to see the prototypes developed. Ravi's boss feels that he should stay back and take care of the presentation since he himself has designed most of the products and is the backbone of the team. He even thinks that with his impressive persuasion he might be able to increase the budgetary allocations for the department. Ravi, however, feels that he must go because the team can take care of the presentation. His wife too has arranged for leave in her office and it will not be possible to postpone it for a later time.

SUMMARY

- The IT age reinforces the continuing demand for proficiency in English communication skills from most of its initiators, mid-level force and end-users.
- Communication is a complicated process of give-and-take with innumerable intricacies and dimensions.
- For communication, more than language, what is needed is an attitude, a willingness to give and take; to open up to others and accept others; to have empathy and a capacity to look at situations from varied perspectives.
- Creativity in day-to-day communication is one's ability to break out of stereotyped reactions, to see beyond the set and accepted patterns.
- Empathy is a quality that will allow one to understand the mental and psychological state of the person we are communicating with.
- The two dimensions of communication, according to the Johari window are, exposure and feedback. These can be looked at through four windows—open, blind, hidden and unknown.
- It is always desirable to expand the size of the 'open window', increase self-disclosure and be more willing to listen to feedback from others about oneself.
- Passing communication, factual communication, thoughts and ideas, feelings and peak communication are the five levels of communications based on the nature, scope and depth of interaction.

- Much of effective communication is non-verbal. Body language is an important factor of the non-verbal communication.
- Negotiation depends on the nature of exchange one proposes or accepts. It depends on the role one has fixed for oneself and the degree of empowerment that has been negotiated and agreed upon.
- The roles people take during interaction can be that of the parent, the adult or the child.
- Effective interpersonal communication depends on appropriate handling of these three roles.

REVIEW QUESTIONS

1. What is Information Technology?
2. What are the effects of MIS on organizations?
3. What is communication?
4. What factors are necessary for effective communication?
5. What is creativity in the context of day-to-day communication?
6. What is the function of language?
7. What are the elements of human communication?
8. What is empathy?
9. What do you understand by the term 'Johari window'?
10. What are the implications of the Johari window?
11. What are the different levels of communication?
12. How are these levels of communication related to interaction?
13. How is the nature of a specific interaction determined?

PART 1

Grammar Matters

CHAPTER 1

Tenses, the Active and the Passive Voice, and Reported Speech

> *Like everything metaphysical,*
> *the harmony between thought and reality*
> *is to be found in*
> *the grammar of the language.*
>
> —*Ludwig Wittgenstein*

Chapter Objectives

This section aims to give the students the basics of grammar. It deals with tenses, the active and the passive voice and the reported speech. It also deals with the main and auxiliary verbs, the finite and the non-finite verbs and the way they are used in the different tense forms. The chapter deals with grammar within very limited space. Hence, only some of the very important concepts have been dealt with.

TENSES

The verb The verb is one of the most important elements of English grammar. It can be described as a word that expresses action in a sentence. Look at the following expressions.

a. You wait patiently.
b. You wait.
c. Wait patiently.
d. Wait.
e. patiently
f. you

The first four here can be seen as complete expressions. They can stand alone. The last two, however, cannot be independent. The main difference between them, if you see closely, is that the last two do not have a verb. It is the verb that forms the nucleus of a sentence. Hence, it makes the sentence meaningful. We can have a sentence with only the verb but there cannot be any sentence without a verb.

The Main and the Auxiliary Verbs

Now, look at the following sentences:

a. *Children play happily.*
b. *We can do the work in three days.*
c. *The workers have been working overtime.*
d. *The typed papers had been lying on the table for months when the enquiry was made.*

In the first sentence, the verb is "*play*". It expresses the action in the sentence. But in the second sentence, the verbs are, "*can*" and "*do*". And in the third sentence, "*have*", "*been*" and "*working*' are verbs. These verbs, however, are not similar. They are either the main or the auxiliary verb. Generally, words like "play", "do" or "work" are the main verbs. And words like "can", "have", "been", etc, are the auxiliary verbs. It could be said that words that express the action in a sentence are the main verbs while verbs that help determine the time frame are auxiliary verbs. The following are the categories of words we use as auxiliaries.

"Modal" forms -- can, could; will, would; shall, should; may, might; must, ought to.

"Be" forms – is, am, was were, are.

"Have" forms—have, has, had.

Now, change the tense of the sentences given above:

a. *Children played happily.*
b. *We could do the work in three days.*
c. *The workers had been working overtime.*
d. *The typed papers have been lying on the table.*

We find in each of the cases that one of the words identified as the auxiliary verb changes the form in case of a tense change. But in case there is no auxiliary verb, the main verb changes its form (as in the case of the first sentence).

A sentence will always have only one main verb. But it can have one or more than one auxiliary verbs.

Look at some more sentences:

a. *Man is mortal.*
b. *This city has an efficient government.*
c. *All children are creative by nature.*
d. *Dinosaurs were both vegetarian and non-vegetarian.*

These sentences have the words we have earlier identified as auxiliary verbs. But as you see, they are the only verbs in the sentence. Hence, they are the main verbs in these sentences. One proof of this is that, in case of a change of tense too, these forms will undergo change.

ACTIVITY

1. Look at the following sentences and identify the main and the auxiliary verbs in each. Point out when the main verb is changing its tense.

a. The identities of the people were not revealed.
b. The glaciers in the north and the south pole are melting fast.
c. The atmosphere is getting polluted very fast.
d. The rare birds had left their habitat before the team of experts could reach them.
e. I have broken my hand.
f. This building has stood the test of time.
g. We have been waiting for the bus since five o' clock.

The Finite and the Non-Finite Verb

One more distinction you should keep in mind is that between the finite and the non-finite verb. The verb which carries the tense in a sentence is always the Finite verb and the others are the non-finite verbs. Look at the following pairs of sentences—

a. *the grass on the other side <u>is</u> always green.*
b. *the grass on the other side <u>was</u> always green.*
c. *the grass on the other side <u>has</u> been always green.*
d. *the grass on the other side <u>had</u> been always green until we crossed over and <u>saw</u> it for ourselves.*

We find that in the first pair, there is only one verb and naturally, it caries the tense. But in the second pair, there is more than one verb, and when there is an auxiliary verb, it is the auxiliary that caries the tense. In a sentence, the verb which carries the tense is called the finite verb and the others are called the non-finite verbs. If a sentence has only one verb, the main verb is the finite verb. But if there is more than one verb, the verb carrying the tense is the finite and the rest are non-finite verbs. It is necessary for you to be clear about these two types of verbs to master the tenses that follow in the later sections.

The Forms of the Verb

Look at the following sentences:

a. *We walk to the ground every morning.*
b. *The bus generally arrives on time.*
c. *The children played the whole day yesterday.*
d. *The executive members are planning for a meeting next month.*
e. *The students have seen the exhibition.*

These five sentences use the five forms of the verbs that are used in different ways to express the tenses. The five forms are the following:

a. The base form: walk
b. The –es form (the third person singular): arrives
c. The past tense form: played
d. The be –ing form: planning
e. The have –en form: seen

These are the five forms of the verb in English. It is important to note that these are the different grammatical forms of tenses. The tenses in English do not necessarily coordinate with the times they express. Hence, it is important to see how these different forms express the different time sequences. Given in the units that follow, are descriptions of the tense form and the time sequences they express.

> The verb "be" has five forms—"**is**" "**am**" "**are**" "**were**" "**was**". The first three are used for the present tense form and the last two for the past tense form.

The Simple Present Form

> Look at the following sentences:
> *Rammohan drives the buses in this route.*
> *Mahanadi flows into the Bay of Bengal.*
> *We experience the effects of ozone depletion every day.*
> *He/she feels it is good in the long run.*
> *They/we feel it is good in the long run.*

The simple present tense, we have seen, uses two forms—the third person singular *–es* form and the *base* form. We use it to indicate things that happen all the time and to show universal truths:

a. *The earth goes around the sun.*
b. *The moon reflects the rays of the sun.*
c. *The flight takes off at 5 pm.*
d. *Continuous working leaves one tired and exhausted.*

Sometimes we also use the simple present to show how often we do things or how often things happen.

a. *Rohit sometimes plays cricket.*
b. *How often do you go to the dentist?*
c. *We usually take the road to the right of the park.*

We use do/does to express questions and negative sentences:

a. *Do you believe in Darwin's theory of evolution?*
b. *Rice doesn't grow in cold countries.*
c. *We don't often do things according to others wishes. Do we?*

Activity

2. Fill in the blanks by using the right form of the verb given in brackets.

a. The shopping mall __________ (open) at 9am everyday.
b. We often __________ (see)
c. I have a jeep but I __________ (do) drive it often.
d. The river Amazon __________ (flow) into the Pacific Ocean.
e. How often __________ (do) you write to your parents?

The Present Continuous Form

Look at the following sentences:
Don't make so much noise. I am studying.
It is not possible to go out now. It is raining.
Have you heard the latest? Ram is finally writing the report.
Can you call the children inside? They are making a lot of noise.

The present continuous uses the "be (is, am, are)—ing" form of the verb. It is used to denote action continuing the present at the moment of speaking.

a. *Talk slowly. The patients are resting.*
b. *We have to go soon. Sheela is leaving.*
c. *Take an umbrella with you. It is raining.*
d. *Be careful. The vehicle is still moving.*

It is also used to show action around the time of speaking.

a. *You are working a lot today. Do you have a deadline to meet?*
b. *The population is rising very fast.*
c. *Your English is getting better. Keep trying.*

Activity

3. Fill in the blanks, using either the simple present or the present continuous(be —ing form of the verbs put in brackets. Indicate which of these situations are temporary and which are permanent.

a. Arundhati is in India now. She ____________ (stay) at the Taj Banjara. She generally ____________ (stay) ____________ there whenever she visits.
b. My parents ____________ (live)in Delhi now. But we are from Bhubaneswar.
c. I ____________ (teach) Physics. But this semester, I ____________ (teach) Mathematics too!
d. I generally ____________ (drive) the bike. But because of a shoulder pain I ____________ (drive) the car these days.
e. He never ____________ (smoke) in public. Looks like something is different today.

We find in these sentences, that the combination of the *simple present* and *present continuous* is also used to show the difference between the permanent and the temporary. There are some verbs in English that can be used in the simple present form only, even when they denote the present continuous action. For example, we can say

a. "I don't understand you." Not, "I am not understanding you."
b. "I believe you." Not, "I am believing you."

The following are some of the verbs that cannot be used in the present continuous form.

want	need	like	love	belong	see	hear
realize	mean	forget	remember	prefer	seem	know

Simple Past

Look at the following sentences:
a. Look! it is raining again.
b. Yea. And it rained the whole of yesterday too!
c. I am hiring a taxi to go to work today. What about you?
d. I usually work from home on such days. Guess I'll do the same.

The simple present generally uses the "–ed" to form the past tense. However, there are many important verbs that are irregular.

a. leave—left *We left for the party at 6 pm yesterday.*
b. go—went *I went to Chennai to see a friend of mine.*
c. cost—cost *This dress cost me a packet.*

Questions and negatives in the simple present use **did/ didn't**.

a. *We walked home.*
b. *Did you walk home?*
c. *They didn't walk home.*

Activities

4. A friend of yours has just come back from a holiday. You are asking her about it. Make complete sentences, using the clues given.

a. stay/cousin ____________
b. what/place/see ____________
c. meet/old friends ____________
d. visit /library ____________

5. Fill in the blanks, using past tense forms.

a. Rakesh ____________ (not/bath) this morning before going to work. He simply ____________ (not/have) time.
b. After the party, we ____________ (decide) not to eat anything because we ____________ (be) not hungry.
c. She ____________ (be) not interested in the book because she ____________ (not/understand) it.

Past Continuous

Look at the following sentences:
a. I saw Ramesh in the garden. He *was reading* a book.
b. We could not go out yesterday. It *was raining* all night.
c. Tom burnt his hand when he *was cooking* dinner

Like the present continuous, the past continuous too uses the "be—ing" form. But while the former uses the "is", "am", "are" (the present tense forms) , the latter uses "was" and "were"(the past tense forms).

The past continuous is used to show action that continued for some time in the past. look at the following sentences.

a. *The child is tired. She was playing all day.*
b. *I have seen the plane. It was flying very low in this area.*

The past continuous is also used to show that some action was continuing when something else happened and intervened.

a. *I was working when the door bell rang.*
b. *We were walking when the car hit us from behind.*
c. *Mohan was cooking when I reached home.*

Activities

6. Make sentences using the words in brackets. Use them in the simple past or the past continuous form.

a. (phone/ring/have/a shower)
My phone ____________ .

b. (watch/a film/television/news flashed)
We were ________________________ .

c. (sleep/lightening/struck)
I was ________________________ .

d. (discuss/ the issue/ director/walk in)
We were ________________________ .

7 Fill in the blanks using either the simple past or the past continuous.

a. Raghu ______________ (fall) off the ladder while he ______________ (paint) the roof.
b. Smita ______________ (take) a photograph when I ______________ (not/look).
c. I ______________ (break) a plate last night when I ______________ (wash).
d. Susan ______________ (wait) for me when I ______________ (arrive).

Present Perfect

Look at the following sentences:
a. Sita *is looking* for her keys. She cannot find it.
b. She *has lost* the keys.
c. I *have seen* the film. I am not really interested in seeing it again.

The present perfect tense uses the "have –en" form, or the past participle form with "have/ has". It is used to show action that happened in the past but is connected with the present.

a. *I have lost my papers.* (*I am still searching for them.*)
b. *Sonia has gone to Delhi.* (*She is still in Delhi.*)
c. *The team has arrived.* (*They are still here.*)

Contrast this with the simple past tense.

a. I lost my papers. (But I managed the situation somehow)
b. Sonia went to Delhi. (She has come back last week)
c. The team arrived. (But they had to leave for security reasons)

We see in these sentences that both the present perfect and the simple past show past action. But while present perfect suggests that the impact of the past action is still there in the present, the simple past might simply focus on the fact that the action is over.

The simple present is often used with "just" to show that the action in question has just been completed.

"Sheela **has just completed** the assignment."

ACTIVITY

8. This exercise has situations. Read them and write a suitable sentence using the verb given in brackets.

   a. Rupa's table was dirty. Now it is clean. (arrange)
      She ________________________________.

   b. Yesterday Sapan was playing cricket. But today his leg is in a cast. (break)
      He ________________________________.

   c. The car has stopped. There is no petrol in it. (run out of)
      The car has ________________________________.

The present perfect is also used to show that you have not done something during a period of time which continues up to the present.

a. *I have never been to the 3D theatre.*
b. *I haven't smoked for a year now.*
c. *I haven't acted in a play since last September.*
d. *Suman hasn't written to me for a year now.*

> Often, we use "for" and "since" with the present perfect tense.
> "For" is used to show a period of time.
> "Since" is used to show a point of time when the action began.

Present perfect is used with expressions like **this morning/this evening/today/this week/this term** etc., especially when the period is not finished at the time of speaking.

a. I have written ten pages since today morning.
b. We have seen five accidents this week.
c. Sheela has taken twelve classes a week this term.
d. We have driven thirty kilometers this evening.

The present perfect is also called the tense of indefinite time. It cannot be used with the mention of specific time. So when a specific time has to be mentioned, we use the simple past instead. Look at the following examples:

I ate out with my friends last week (**not**)
I have eaten out with my friends last week.

I saw the movie last week (**not**)
I have seen the movie last week.

I spoke to Meena on Wednesday (**not**)
I have spoken to Meena on Wednesday.

Activity

9. Answer the questions given, and fill in the blanks using the clues.

   a. Have you seen the new canteen?
   I ______________________ yet, but I'm ______________________.

   b. Have you been to the new block of the department?
   I ______________________ once, but I ______________________ seen it for ______________________ now.

   c. Have you bought your new car?
   Oh yes! I have it with me since ______________________.

Present Perfect Continuous

Look at the following sentences:

a. i. Has the bus come?
   ii. No, we all have been waiting.

b. i. The ground is wet. Did it rain?
   ii. Yes, it has been raining since morning.

The present perfect continuous uses both the "have –en" and the "be—ing" form. It is generally used to show action that began and has continued for some time in the past. These are actions that might have just got over or still continuing at the point of speaking. Look at some more examples:

a. *You are out of breath. Have you been running?*
b. *I have been talking to your teachers about your problem. They feel that............*
c. *The roads are flooded. It has been raining for four hours now.*
d. *I have been watching the TV since five O'clock.*
e. *Aren't you tired? You have been talking the whole evening.*

Now, look at the following sentences:

a. *Ranga's clothes are soiled. He has been painting the ceiling.*
b. *The ceiling was white. Ranga has painted it.*

In the first sentence here, the emphasis is on the action itself. But the second emphasizes on the completion of action.

a. *There are dark circles around Rekha's eyes. She has been working hard to complete the book.*
b. *Rekha has finished working on the book.*

In the first sentence here, the emphasis is on the act of Rekha's working on the book. But in the second, the emphasis is on the completion of the action.

The present perfect continuous, thus, is used when we want to stress on the act of doing something. And only the present perfect is used when it is necessary to highlight the completion of the action.

You can be a little ungrammatical if you come from the right part of the country.
—Robert Frost

Activities

10. In the sentences given below, fill in the blanks by putting the verbs in the correct form — present perfect or the present perfect continuous.

    a. Look! Somebody ________________ (spoil) the freshly painted wall.
    b. I smell cooked food. Have you ________________ (cook).
    c. Shekhar is an actor. He ________________ (appear) in several films.
    d. I ________________ (read) the book you gave me. But I ________________ (finish) it yet.
    e. I ________________ (loose) my key. Can you help me look for it?

11. Given below are some situations. Read them and make sentences using the words given. Use the have –en form for this.

    a. Rupa is from Maharastra. Now she is traveling all over India to raise funds for charity.
    Rupa, from Maharastra, ________________ (traveling) all over India.
    b. Jagan is a tennis champion. He began playing when he was 11. He is playing for India still.
    Jagan ________________ tennis since he was 11.
    c. Jane and Ditty started making advertisements when they were in college. They are still working together on several projects.
    Jane and Ditty ________________ (work) together since their college days.

Past Perfect

Look at the following sentences:

a. *When I arrived at the party, most of the guests had left.*
b. *By the time we found the umbrella, the rain had stopped.*
c. *The car broke down yesterday. The brakes had failed earlier.*

The past perfect tense uses the past form (had) of the "have—en" form to make sentences. In the first sentence here, we find two time sequences. One is *"arriving at the party"*, and the other is *"leaving of the guests"*. If *"arriving at the party"* is a past, the *"leaving of the guests"* happens before this past. The past of the past here uses the past perfect tense. Hence, it is the past of the past. The same is found in the other two sentences too. In all these sentences, there are two time sequences. One—the past, and the other—the past of the past. The past perfect is used to show the past of the past.

Look at some more examples:

a. i. Was Sheela there when you reached the spot?
   ii. No. she had already left.

b. i. Could the police catch the thief yesterday?
   ii. They tried their best but thief had left much before they could arrive.

Activity

12. Fill in the blanks using the right form of the verbs given below.

    a. Jim was not at home when I reached.
    He ________________ (already/leave)

    b. The man was a complete stranger to me.
    I ________________ (never/ see/before)

    c. I was very tired when I reached home.
    I ________________ (play/tennis/wholemorning)

    d. Mr Mathur no longer had his car.
    He ________________ (sell/last month)

    d. I invited Margaret for dinner. But she could not come.
    She ________________ (promise/someone else)

The Future

In English, the future tense does not have a specific form. It is generally expressed by using a combination of various other tenses. Some of the ways of expressing the future time are as follows:

By using the present continuous form with a future word.

a. *I am going to meet the principal tomorrow.*
b. *The plane is leaving in an hour's time.*
c. *We have to increase the speed. The shops are closing in an hour.*

By using the modals to show possibility or probability in future.

a. *It might rain in the evening.*
b. *The train should reach in two hours time.*
c. *I shall write the final exams this year.*

Both "will" and "is going to" express action one intends to do in future. But with a difference.

— *will* expresses impulsive decision, taken on the spot.

I will see to it that we don't fail again.

— *is going to* expresses scheduled future action.

We are going to write to the Principal about this.

ACTIVITIES

13. In these sentences you have to talk about your future plans. Use the right words to answer these questions.

    a. i. Which car are you going to buy?
    ii. I am not sure. I ________________ (Maruti Suzuki)

    b. i. Is Raghu coming with us to the picnic tomorrow?
    ii. He said he would try to. He ________________ (even/bring/his sister) along with him.

    c. The computer is troubling me a lot. I ________________ (take it/workshop.)

    d. i. Where are you going to be transferred to?
    ii. I ________________ (have to/shift/Bangalore).

14. Fill in the blanks in the following sentences.

    a. I wanted to be the first one to give the news. But I was too late. Someone ________________ her. (already/tell).
    b. I couldn't open the door because someone ________________ it. (lock)
    c. Seema and Hari ________________. I could hear them.(argue)
    d. I ________________ about this for some time now. (know)
    e. I signed the register and ________________ to my room. (go)
    f. I couldn't take the bike because Ravi ________________ it. (use)
    g. When he warned them about the police, they ________________ the country. (leave)
    h. I think I ________________ him somewhere before. (see)
    i. when I reached her room, she ________________ the assignment. (do)
    j. We ________________ for hours when I suddenly realized that something was wrong with the engine. (drive)

THE PASSIVE

Look at the following sentences:

a. Rajiv has completed the experiment.
b. *The experiment has been completed.*

a. Many of the residents saw the flying saucer.
b. *The flying saucer was seen by many.*

In the passive form of the sentence, it is the object of the action that is highlighted rather than the agent or instrument of action.

In the passive form, we use the correct form of the "**be**" verb (is/are/were/was) along with the **past participle** form of the verb in the sentence. So we have forms like:

a. the work was done ...
b. the room was cleaned …
c. the house was built …
d. the infiltrators were seen …

Apart from the "be" form, the passive also uses the "**have / has been** ...." form. We have forms like:

a. My bicycle must have been stolen.
b. All the trees in the area have been cut.
c. The monitor has been repaired.
d. The whole class has been punished.

Compare this to the active forms of the sentences:

a. The boys next door must have stolen my bicycle.
b. The illegal traders have cut all the trees in the area.
c. The new service engineer has repaired the monitor.
d. The Principal has punished the whole class.

As mentioned before, the active forms of the sentences focus on the agent of the action—*who* has done the action is as important or sometimes more important than the action. In the passive however, the action, not the doer, becomes important.

Activity

15. Fill in the blanks, using verbs of your choice in the appropriate passive form.

a. No decision on this matter can ________________ till next morning.
b. The book will have to ________________ as soon as possible.
c. The injured man ________________ to the hospital.
d. It is customary for the luggage to ________________ by the customs officials.
e. We all wanted to ________________ before day break tomorrow.

The Passive is used when:

a. The doer of the action is not known.
b. The doer of the action is not important
c. The doer of the action prefers not to be known.

Forms of the Passive

The following are some of the commonly used forms of the passive:

Simple present:

a. The rooms are cleaned everyday
b. People are injured in accidents everyday.
(Everyday, routine action.)

Simple past:

a. The documents were destroyed.
b. Patients were neglected.
(Action in the past.)

Present continuous:

a. Shops are being closed down everyday.
b. Problems in policy implementation are being discussed in detail.
(Action taking place at the time or around the time of speaking.)

Past continuous:

a. We suddenly realized that we were being followed.
b. Different ways of lessening the border tension were being discussed when the war broke out.
(Action that continued for some time in the past.)

Present perfect:

a. All of us have been invited to the party
b. Shyama has been elected president.
(Action was in the past time but it is still valid at the time of speaking.)

Past perfect:

a. The rooms looked much better. They had been cleaned.
b. We simply could not voice our protest. We had been asked to keep quiet.
(Action that took place before the past action mentioned in the sentences.)

Activity

16. Given below are some sentences. Read them and write down another sentence with the same meaning. begin the sentence the way it has been suggested and note whether you are using the active or the passive voice.

    a. All the students should enter their suggestions in the register kept in the office.
    Suggestions ____.

    b. The Principal postponed the meeting due to his ill health.
    The meeting ____.

c. Somebody might have issued the book if it was in the library.
   The book ______________________________________.

d. A short circuit might have caused the breakdown.
   The breakdown ______________________________________.

e. The architects are redoing the office. It cannot be used right now.
   The office ______________________________________.

> *Grammar is a piano I play by ear. All I know about grammar is its power.*
>
> *—Joan Didion*

REPORTED SPEECH

Look at the following sentences:

a. i. "I am going home", he says
   ii. *He says that he is going home*

b. i. They said, "we are going on a picnic"
   ii. *They said that they were going on a picnic.*

Direct speech is the actual speech of a speaker reported without any intervention. This is indicated by putting the actual words within inverted commas, and separating it from the main sentence with a comma. (see the sentences given above in the box). The reporting of this direct speech to someone else needs a slight alteration of structure. This is called **Reported speech**.

Look at some more examples:

Charlie said, "I am thinking of going to Canada."
Charlie said that he was thinking of going to Canada.

Raghu said, "I have been playing a lot of tennis."
Raghu said that he has been playing a lot of tennis.

Sheela said, "I don't know what Rajiv is doing."
Sheela said that she did not know what Rajiv was doing.

Reported speech, we find here, introduces a "that…" form and changes the tense of the reported clause. But this does not always happen. Different tense forms make their reported speeches differently. Remember that in a sentence where a speech is being quoted or reported, there are two clauses with two finite verbs – one is the main verb that is linked with who did the saying and the other talks of what was said. Given below are some verb forms and the way they are transformed into the reported speech.

- When the main verb of the sentence is in the present, the present perfect or the future tense, there is no change in the reported statement.

"I haven't done my homework," she says.
She says that she hasn't done her homework.

"The plane will land in fifteen minutes," the pilot has just announced.
The pilot has just announced that the plane will land in fifteen minutes.

- When the main verb of the sentence is in the past tense, the verb of the reported speech also is changed into the past. Look at the following sentences:
  a. "The washing machine is broken," he said
  He said that the washing machine was broken.
  b. "We have never been to Berlin," they said.
  They said that they had never been to Berlin.
  c. "I am not feeling very well," Shilpa said.
  Shilpa said that she was not feeling very well.
  d. "It was raining very heavily yesterday," she said.
  She said that it had rained very heavily that day before.

The following are the tense changes we find when direct speech is converted to reported speech:

simple present	*simple past*
present continuous	*past continuous*
past continuous	*past perfect continuous*
present perfect	*past perfect*
simple past	*past perfect*

While changing the direct speech into the reported form the following changes too are necessary.

shall	should
will	would
must	had to
can	could
tomorrow	the next day
yesterday	the day before
today	that day
here	there
this/that	the

Activity

17. Change these sentences into reported speech, changing the words where necessary.
  a. "I'm listening to the radio," he said.
  b. "I stayed awake the whole night after the party," she said.
  c. "We can phone home once we reach the airport," they said.

d. "I have to see you tomorrow," my project leader said.
e. "I met her about three months ago," the stranger told.
f. "Rajiv is bringing some new records to the party tonight," Suraj said.
g. She said, "we went swimming today."

Reported speech and "wh" words

When sentences have a "wh" word, they change their word order when changed into reported speech. Look at the following sentences:

"What's the time?" she asked.
She asked what time it was.

"How's your mother"? He asked.
He asked me how my mother was.

Reported speech and commands

Reported commands also undergo change. They can use a number of verbs like "order", "command", "warn', "instruct'.

"Don't get worried", she told.
She requested them not to get worried.

"Finish the job tonight please", my boss told me.
My boss requested me to finish the job tonight.

Activity

18. Read the sentences given in direct speech and make them into reported speech with similar meanings.
    a. "Listen carefully to the music," he told us all.
    b. "Don't you ever come before 6 o'clock," she told him sternly.
    c. "Open the door please," my mother told me.
    d. "My parents are arriving tomorrow," she said.
    e. "We visited her this morning," the nurse told.
    f. "I am going to visit the Unites States next year," the PM announced.
    g. "I don't smoke," Rekha said.
    h. "I haven't seen my mother for years," Zeenat said.
    i. "It is difficult for me focus on things," my mother says.
    j. "It is difficult to escape one's fate," it has been told.

SUMMARY

- Tenses express the action in a sentence.
- The main verb gives the content of the action, while the auxiliary verb gives the appropriate framework of the action.
- The finite verb carries the tense in a sentence.
- The "be" verb has five forms—is, am, are, were, was.
- The simple present and the present continuous when juxtaposed, show the difference between the permanent and temporary action.
- The present perfect does not express definite time.
- The passive voice highlights the action or the process involved, whereas the active voice emphasizes on the agent or instrument of action.
- Passive voice is used when the doer is either unknown, unimportant or obvious.
- Reported speech is used when there is the need to report what has been stated indirectly.
- The tense of the reported speech undergoes change only when the main verb is in the past tense.

PART II

Communication Matters

CHAPTER 2

Non-Verbal Communication: Body Language

Get in touch with the way the other person feels. Feelings are 55% body language, 38% tone and 7% words

—Anon

Chapter Objectives

This chapter introduces the subject of body language. It discusses the communicating potential of our body along with its gestures. It looks especially into the communicating ability of the face, hands and the feet. The chapter also examines the concepts of distance and positioning, orientation of the body and mirror imaging.

THE COMMUNICATING BODY

Communication, as we have seen before, can be both verbal and non-verbal. If verbal communication deals with words, sentences or spoken expressions, non-verbal language constitutes body-movements, gestures and facial expressions. In fact, research has proved that only 7% of our communication takes place through words. About 38% of the message is conveyed through the tone, voice and inflection while and non-verbal, physical behaviour accounts for around 55% of our communication. Very often, body language is unintentional and involuntary. Ironically, it is these signals that primarily determine the quality of communication, the meaning that is finally read into what we want to say.

Body language forms a very important part of our day-to-day communication. They express our inner feelings, our conflicts and interpersonal attitudes. Often, when we fail to express our feelings in words, a warm handshake or a hug, conveys intimacy much more effectively and powerfully. Similarly, loss of eye contact, yawning, or slumping in the chair can express boredom or loss of interest, even though verbally we still say that we are interested.

> *Too often we underestimate the power of a touch, a smile, a kind word, a listening ear, an honest compliment, or the smallest act of caring, all of which have the potential to turn a life around.*
>
> —*Leo Buscaglia*

STUDYING BODY LANGUAGE

Discussed below, are three widely studied categories of body language. They are: 1) The facial expressions, 2) Movements and positioning of the hands, 3). Movements and positioning of the legs.

The Face

In any communication situation, the face is generally said to be the most effective indicator of emotions. Most of our reactions as a speaker are dependent on the feedback we get from the face of the listener. Many professionals who directly deal with people like teachers, counselors, doctors, businessmen and artists take a lot of clue from the faces of the person they are dealing with. Researchers have found generally eight basic facial emotions: surprise, interest, joy, rage, fear, disgust, shame and anguish. These are universally found in all cultures. Sometimes, the face may even show a blend of two different emotions like, for example, pleasant surprise, which can manifest as a pleasant smile combined with raised eyebrows.

Given below is a list of the facial expressions and the emotions they suggest.

a. Surprise or astonishment	- Raised eyebrows.
b. Anger or frustration	- Furrowed forehead.
c. Displeasure or confusion	- Frowning.
d. Antagonism	- Tightening of the jaw muscles or squinting of the eyes.
e. Defiance	- Thrusting out of the chin.
f. Dislike, rejection or contempt	- Turning up of the nose.
g. Extreme shock	- An open-mouthed reaction. (This can also happen if someone is concentrating very intensely. In this case, all the facial muscles below the eyes are completely relaxed and sometimes even the tongue protrudes).
h. Aggression	- Here, the eyes will be wide-open and the lips tightly closed. The corners of the eyebrows will be turned down and teeth clenched.

i. Disinterest in a deal - Eyes down-cast with the face turned away.

j. Interest - The opposite can be a relaxed mouth, a smile and a projected chin. This can suggest that a customer is considering a deal and is obviously interested.

k. The blank face - This is the 'dead pan' face or the emotionless face that is completely relaxed. This is a typical reaction when a person is withdrawn, completely into himself or herself. The strong emotional signal that this face sends is, "Do not disturb".

Most of the facial reactions we have discussed here are a combination of two or more features. Sometimes a single aspect of the facial expression too can convey a lot.

The head nod: A small head nod generally signifies attention whereas a more vigorous and repeated movement signifies strong agreement with some viewpoint being expressed.

Head movements are generally very significant in communication. If a person shakes his/her head from side the side but says—"Of course I do agree" or "This is really interesting" one cannot definitely take the words at their face value. The feeling one is left with is that, either the speaker in lying or being sarcastic.

The head tilt: Slight tilt of the listener's head is always seen as a sign of keen listening and interest. The nodding of the head accompanied by the head-tilt position can suggest empathy and complete agreement with the speaker. In contrast, the straight head, more often than not, shows a neutral attitude.

During a lecture session, if heads are not tilted it might be an indication that sufficient interest has not been evoked. If from a head tilt position people move on to a straight position and then begin to slouch, it is a clear indication that there has been too much of information loading and that it is now time to stop.

Head tilt

Open hands

The Hands

Our hands are among the most expressive parts of our body during communication. In a big way, they indicate our mental process. Scientists have observed that there is more number of links between our brain and our hands than any other part of the body. They signal our thoughts, show our feelings and our mental state.

Open hands: Open hands are always a sign of trust, a desire to communicate and an invitation to share his/her point of view. Some even believe that it allows positive energy to flow from the speaker to the listeners, making the listeners more receptive to what the speaker is saying.

Upfacing flat hand: A flat hand with the palm facing up or outward is always definitely a silent question. It says, 'why?' or "I don't understand'

Relaxed hands: A calm, confident, self-assured person's hand will definitely be relaxed. The hands would gracefully rest one on top of the other. Even though they are involved in doing something else, their movement will be steady and controlled.

Relaxed hands

Clenched hands

Restless hands: An uneasy, jittery or nervous person's hands will also be restless and jittery. Even when apparently calm, the hands can be restlessly picking, scribbling, biting or even sucking. Such movements show anxiety or a lot of nervous energy.

Clenched hands: Invariably, clenched hands are a sign of negative emotion. In different contexts, they signify different emotions. Sometimes, they signify a feeling of tension and frustration (a feeling of being tied down). During negotiations, they can even signify a closed mind, a persons who will no longer accept suggestions or see your point of view. Clenched hands with the rubbing of the thumb or picking the cuticle of one thumb, sometimes shows need for reassurance. They signify a nervous, uncertain mind, unable to resolve a problem.

Clinging hands: Sometimes, when we cling to the armrest of the chair, a table, files or books, we indicate insecurity, or nervous anxiety.

Wringing hands: Pressing, twisting and squeezing of one hand with the other is again a sure sign of discomfort, feeling of insecurity or need for reassurance.

The Arms

Crossed Arms: Crossed arms form a kind of barrier, a guard against perceived threat. It can be seen as a defensive posture. Sometimes, it may even mean that the person being spoken to has a closed mind to the topic being spoken. He/she is determined to take up a certain stance and is subconsciously bracing up against any opposition.

Crossed arms

Arm gripping

The gesture also stands for relaxation. It can be seen as a way of resting the arms. However, generally it has also been seen that the person who takes up this posture is less likely to listen or accept the viewpoint of others. In a group, generally, people who have their arms relaxed by their sides are more likely to be ready to listen, accept or understand.

Arm gripping: Arm gripping, similarly is seen as a negative attitude. In this posture the hands grip the upper arms tightly to prevent their sliding down to a relaxed position. More often than not, this posture shows a negative, suppressed attitude. It signifies fear or sometimes depression. This also is a subconscious protective posture the body assumes when there is any psychological threat or perception of insecurity.

Arms behind the back: If a person tightly clenches his fist behind his back or tightly holds the wrist of one hand, it either shows that he/she is trying to disguise emotions or facing inner conflict. The same gesture – arms at the back – but with unclenched hands can, on the other hand, show superiority and confidence.

Unconsciously, here, the person has exposed his vulnerable areas like throat, heart and stomach. The gesture thus shows fearlessness, authority and easy confidence.

Apart from these basic kinds of hand and arm movement, there are numerous hand-to-face gestures that signify certain moods or attitudes.

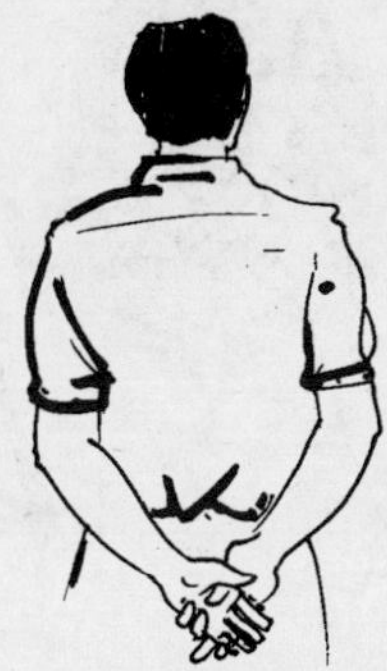

Arms behind the back

Hand covering mouth

Hand covering the mouth: This gesture might show that the person is lying, trying to hide a feeling, or deliberately not expressing his views.

Putting something in the mouth: This is done in many ways. A person may bite his nails, put pens, pencils or even the arms of spectacles into his mouth. All these are seen as movements that show the need of reassurance. They can indicate anything from nervousness, indecisiveness and anxiety to a decision-delaying tactic.

Putting something in the mouth

Scratching the neck

Scratching the neck: A sustained scratch accompanied by the person looking down or away could mean that he/she is searching his/her memory. Scratching the neck just below the ear lobe lightly, would mean uncertainty. It shows discomfort.

Neck stroking: This is more often found in women. They stroke the neck or feel the necklace. It would mean disbelief, discomfort or even signify indecisiveness or a decision making process.

Rubbing the eye: Shutting the eye with a finger or rubbing the eyes during a conversation could mean that the person is in doubt or is trying to deceive.

Rubbing the eyes

Rubbing the nose

Rubbing the nose: Rubbing or touching the nose slightly during a conversation might be a sign of rejection or a gesture of doubt.

Hand behind the back of the head: This is a gesture that signifies confidence, dominance or a feeling of superiority. Often people are found assuming this posture during conversation with their subordinates but is done rarely in the presence of their superiors. It is a typical gesture that means "Everything is under control".

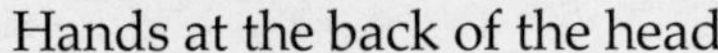

Hands at the back of the head

Hands on hips

Hands on hips: This is a gesture that could show aggressiveness or a non-verbal challenge. Very often,however, it shows readiness for action, readiness to take up or tackle targets. Sometimes, people adopt this posture to appear bigger, move powerful or dominant.

The Legs

The placing and positioning of the legs too, play an important role in body language. Given below are some of the big postures we commonly make.

Cross-leg gesture: This is a gesture when the legs are crossed, with one knee resting on the thigh of the other. It is generally done when a person is occupying a chair, sitting in the same uncomfortable position for a long time. Variation of this posture is one leg crossed almost horizontally, with the ankle resting on the other knee. Like the crossed hand position, both these positions show a defensive attitude. The person might be nervous, withdrawn, or geared up to face a perceived threat. It is generally observed in meetings that when there is a discussion or negotiation many of the participants will cross their legs. When an agreement is reached, and the issue has been amicably settled, many of them will most probably uncross their legs and move or tilt towards one another. Generally, it may be concluded that crossed legs signify defensiveness, a propensity to not accept the other viewpoint. Uncrossed legs, on the other hand, show a more receptive frame of mind, in tune with what the speaker is proposing. If the crossed-legged position is coupled with the crossed hand position, the resistance is definitely intense. In a meeting or during discussions, this is most surely a sign that there is an argument or resistance ahead.

Crossed legs with the foot moving in small kicking motion, again, signify boredom in a given situation. This, combined with the foot pointing towards the exit is a clear sign that the person wants to leave the place.

While analyzing crossed legs, however, one has to be a little careful. In Europe, for example, women are expected to sit with their legs crossed or tightly held together. Etiquette demands it. In this context, therefore, it need not be inferred that the women are defensive or in a mood to disagree.

Crossed legs

Relaxed sitting pose

Sitting Styles

The relaxed sitting pose: Sitting in a chair with legs relaxed and apart, the hands resting flat on the arms of the chair signifies an open, relaxed attitude. Two people sitting in this posture facing one another and discussing, is definitely a sign that the two are in tune with one another's thought. There is hardly any disagreement or resistance.

Sitting on the edge of the chair: From a relaxed sitting posture, moving to the front of the chair can signify that the people involved are ready to compromise, co-operate or agree. Sometimes, it can even signify that the people are ready to finally reject and abandon the discussion. Sitting forward in a chair with feet on the tiptoe shows readiness for acceptance, readiness for action.

Sometimes, when sitting on the edge of the chair is accompanied by stiff hands placed on the knees or gripping the sides of the chair, it can signify fear, tension or a desire to be free from the situation. This is generally coupled with very tense facial expression.

Leg over the arm of the chair: Apparently, a person with the leg over the arm of the chair appears relaxed and informal, in a mood to enjoy. But sometimes,the posture can also be taken to signify a defensive attitude saying '"I don't care!". It can denote indifference, hostility and sometimes a dominant, superior attitude.

Straddling the chair: Sitting with the chair around so that its back faces the people one is talking to can signify an informal, relaxed mood. In a different context, however, it could even signify a defensive mood, an attempt to shield or protect oneself against imagined hostility. A straddle can show aggression, especially when faced with resistance or bored with the conversation. In such a context, turning around and going to the back of the person one is talking to can make the latter feel vulnerable. And most probably he/she will turn around to face the speaker, if it is a swivel chair.

Stradling the chair

DISTANCE AND POSITIONING

In our day-to-day life, very often our attitude is signified by the spaces we occupy – both in relation to others or ourselves. We stand in particular places, occupy particular chairs and maintain specific distances vis-à-vis specific people. Space or distance can, in general, be divided into these broad categories.

Intimate distance: Each of us guards our own territory. We allow only certain people into very close proximity. This could be from real physical contact to about six inches from our body. This distance we reserve only for people with whom we are emotionally close. It could include one's child or parent, close friends, relatives or a lover. Here, the sense of physical proximity is intense. But visibility is low.

Personal distance: This, too, is a relatively close distance zone. It involves a certain area a person reserves around oneself. In case of informal, outgoing people, this distance might be less. Stiff and reserved people however might maintain a larger area as their personal distance. This is typically the space within which a person maintains privacy or social intimacy. This distance becomes obvious in social gatherings or parties. Sometimes two or more people gather in a public place but get together within one another's personal zone. This is in striking contrast to the longer distance which people maintain with others with whom they don't feel particularly close.

Social distance: This is the distance we maintain with people we interact, but don't know very well. It could be a new employee, a colleague or sometimes people staying in the same neighborhood. The distance could be between four to seven feet depending on whether it is a close or distant social contact. People with a firmly established hierarchical relationship also generally maintain this kind of distance.

Public distance: This could be a distance between twelve and twenty five feet -- a distance suitable for lectures or conferences. Such distance is also maintained by politicians while addressing people and by actors on the stage.

BODY ORIENTATION

Sometimes, the way we stand in relation to one another also shows our attitude and frame of mind. Two people with the upper part of the body inclined towards one another definitely means agreement. The distance between the two people could suggest how close or far they are but they definitely agree or at least are tuned to one another's thought process. When two people maintain a right angle and talk without facing one another, it is definitely not a private conversation. The topic could be routine instructions in an office or a discussion that does not closely concern either. When two or three people are discussing something intimate or private they will most probably face one another squarely or form a close triangle. The bodies also might be inclined towards one another. The clear sign here is: "This is a closed meeting–keep off".

Pointing

The direction towards which a person points his feet or his torso is again a very significant pointer to his/her attitude. Feet pointing at the door could suggest a desire to leave. Sometimes, pointing your feet at someone might even suggest that we are interested in the person. In a party, if a man points his foot at a woman, it might mean he is interested in her. She might, in turn, move close to another person signaling her preference.

Mirror Imaging

This is a situation where people mirror or imitate one another's body positioning. After a meeting or during a break if we find two people chatting, mirroring one another's body posture, it is definitely a sign that they agree with one another. Each is receptive to the other's ideas. If in between, one of them suddenly changes his/her position, touches the nose with a finger or looks away, it could be a sign that the rhythm has been broken. Disagreement has cropped in and they are trying to negotiate a problem.

Mirror imaging can be particularly helpful in situations like interviews or a negotiation table. The person whose posture you mirror gets a subconscious feeling of comfort in your presence. It might even make her more inclined to agree with you, accept you and your views. But be careful to do it subtly, delicately. Obvious imitation, which the other person can see and observe, can be seen as a mockery or even as trying to make the other person feel uncomfortable.

Body language is definitely a very important pointer to our attitude, our feelings and emotions. It is important, however, to remember that body language should be considered as a cluster. While studying body language, one should take into consideration the facial expression, the hand, the leg movement and even the context and culture one is operating in. Our experience as social beings does equip each of us to read and recognize body language to a large extent. Often, we have a "feeling" that a person is lying. In spite of talking in a friendly manner, we "think" that he has suppressed hostility towards us. "Instinctively", we know that two people are intimate and we move out of their personal zone. Sometimes, we also "know" that we are being ignored or we are unwanted in a group.

A mother definitely "knows" when a child needs reassurance even though nothing is said. In intimate relationships again, we know exactly whether the person we love is tuned to us or not. Most of this knowledge, we pick up subconsciously from the people around. Some of it could also form part of our genetic memory because even newborn babies have been found to respond to and understand body language. Good communicators, instinctively, understand more of others' body language and respond to it appropriately. It is important for each of us, therefore, to cognize body language as an important part of communication, accept the feedback it gives us continuously and adapt ourselves accordingly.

Activities

1. Given below are some typical behaviour traits in column by A and some expressions that can match them in column by B. Try to identify what expressions can match with the traits listed.

Column A	Column B
1. Turning up the nose	a. I don't believe you. I think it is impossible.
2. Cracking the knuckles	b. I'm energised.I'm ready.
3. Rubbing the nose	c. I want to be honest with you.
4. Sitting on the edge of the chair	d. I want to emphasize this point. I'm determined.
5. Crossing the arms at the chest	e. I dislike this dish.I reject your opinion.
6. Putting the feet on the table	f. I disagree with you. I have to fight it to the end now.
7. Placing the hands on the hips	g. I'm nervous and uncomfortable. I wish I could leave the place.
8. Pointing the index finger	h. I'm bored now.I want to go on to the next item on the agenda.
9. Placing both palms on the chest.	i. I'm indifferent. I don't care about what you are saying.
10. Shrugging the shoulders	j. I'm the boss here

2. After you have matched the columns, build a small monologue around each emotion expressed and enact it

3. Given below is a role-play that involves the expression of various emotions. After the role-play is completed, analyze each of the characters and the range of emotions they expressed. Try to remember the body language and the postures that helped them to express these emotions.

Role-Play

The Situation: This is a joint family where the grand father and the grand mother (Krupa Shankar and Rukmini) stay with their son and daughter-in-law (Aditya and Shanta) and their six-year-old daughter (Priya). The problem has started because, of late, both the father and the mother of the child have started feeling that the grandparents have been pampering the child and that she is getting to become increasingly demanding. There has always been a little tension between the father and son because Aditya felt that his father was very strict in bringing him up and had insisted on his taking up engineering in spite of his desire to take up acting as a profession. The following is a rough description of their characters.

Role Descriptions

Krupa Shankar (grandfather)

You are 68 years old, a retired professor, who always believed in the merits of discipline and hardwork. You sincerely believe that you have done the best for your son.
A career in acting would have been too risky for him and he was not academically motivated enough to get into teaching. In your grand daughter, however, you see a new kind of possibility. With the widening of career possibilities you think she has a great future. You feel she is a brilliant child who should be allowed to choose what she wants in life. Also, you have a nagging guilt in your mind for having stopped your son from taking up his acting career. You would like to compensate by encouraging Priya to follow her heart.

Rukmini (grandmother)

You are almost 65 years old, tired and exhausted with life. You have had an active career and you strongly believe that parents should have the sole deciding authority about their children's future. You are fond of both your son and daughter-in-law, but also detached enough not to be too involved with their life. You feel that Priya should have her own freedom in life, but you don't want to impinge on your son's parenting style.

Aditya (son)

You are 38, a dynamic successful, upcoming engineer. You enjoy all that your present career has given you but you still nurse a secret wound for not being able to take up a career in acting. You resent the fact that your father was once upon a time very strict with you, and you would like to oppose whatever he says or believes. You too think that Priya should have the freedom to be herself, but you want her to be more disciplined and time-conscious. You feel that your father's pampering is encouraging her to be wild and unfocussed.

Shanta (mother)

You are 32, a young lady with your own business enterprise. You feel both your father-in-law and husband are imposing their own ideologies on your daughter. You feel they should just let her enjoy her childhood. Of late, you have been worried about your father's failing health and you are irritated with the constant discourse about your daughter's future.

All of you are sitting together one evening, having tea after a hard day's work. Suddenly, Priya runs in, topples over a cup, breaks it and runs out. Even before you understand, an argument breaks out between father and son. You and your mother-in-law too are dragged into it.

Now analyze the characters along the following lines.

Krupa Shankar

Dominant emotional traits corresponding body language

1.

2.

3.

Rukmini

Dominant emotional traits corresponding body language

1.

2.

3.

Shanta

Dominant emotional traits corresponding body language

1.

2.

3.

Aditya

Dominant emotional traits corresponding body language

1.

2.

3.

SUMMARY

- Non-verbal language constitutes body movements, gestures and facial expressions.
- Non-verbal, physical behaviour accounts for around fifty-five per cent of our communication.
- When we fail to express our feelings, a warm handshake or a hug conveys intimacy much more effectively and powerfully.
- Loss of eye contact, yawning or slumping in the chair can express boredom or loss of interest, even though verbally we may still say we are interested.
- The face is a very powerful indicator of our feelings. Professionals like teachers, counselors, doctors, businessmen and artists take a lot of clue from the faces of the person they are dealing with.
- At times, the face shows a blend of two different emotions, like for example, pleasant surprise. It can show as a pleasant smile combined with raised eyebrows.
- Hands play an important role in communicating feelings. Scientists have observed that there is more number of links between our brain and our hands than any other part of the body.
- The placing and positioning of the legs also play an important role in body language.
- In our day-to-day life, very often, our attitudes are signified by the spaces we occupy—both in relation to ourselves or others around us.
- Intimate distance, personal distance, social distance, and public distance are some of the broad space/distance categories.
- Sometimes, the way we stand in relation to one another also shows our attitude and frame of mind.
- The direction towards which a person points his feet or his torso is a significant pointer to his attitude.
- Mirror imaging goes a long way in making the other person comfortable in our presence and to receive our views positively. It can be particularly helpful in situations like interviews or at the negotiation table.
- Effective communicators instinctively understand more of others' body language and respond to it appropriately.

REVIEW QUESTIONS

1. In a communication situation, what is the ratio of verbal vis-à-vis non-verbal communication? State in short the various factors involved in communication.
2. What are the different kinds of basic emotions expressed by the face? In a situation where a dealer is trying to strike a deal with a client, what are the clues he/she should look for in the face to know the client's reaction?
3. What are the different kinds of head movements that matter during communication? State with examples what each of them signifies.

4. What are the two kinds of hand movements that can differentiate a calm mind from an agitated restless mind? Give examples and show the differences.
5. How would you distinguish between the clenched hand and the clinging hand? In what situation will each of these be used?
6. Give a list of body movements that will signify a calm, relaxed and open mind. Explain each of their characteristics in detail and place them in hypothetical situations.
7. In a group, the way people sit or stand gives away the proximity they have with one another, the interest they have in what the other is saying. Discuss some of these body-movement clusters.
8. Body language should always be studied in a particular context. The same posture, in different contexts can mean different things. Give examples.
9. What are the different kinds of distances we observe during communication situations? State the significance of each.
10. Each of us is naturally endowed with the capacity to read and analyze body language—discuss.
11. What is mirror imaging? State its role in the communication situation.
12. Can we always interpret cross-legged gestures to be defensive gestures? Give reasons for your answer.
13. How many significations does the 'arms behind the back' posture have? State any two with examples.
14. What does 'pointing' mean? What can you infer from a person's feet pointed towards the door?
15. What is the difference between 'scratching the neck' and 'stroking the neck'?

CHAPTER 3

Listening Skills

> *We have two ears and only one tongue in order that we may hear more and speak less.*
>
> —*Diogenes Laertius*

Chapter Objectives

This chapter introduces the mother of all communication–listening. While emphasizing the primacy of listening among all the skills, it distinguishes between hearing and listening. It further discusses the processes of active listening and other kinds of listening. Barriers to listening and good listening are elaborated alongside activities like Chinese whisper etc.

THE LYNCHPIN OF COMMUNICATION

Listening can be described as a skill that involves receiving, interpreting and responding to the message sent by the communicator. Like any other skill, listening skill also needs to be learnt and developed for effective communication. It is, in fact, one of the most important skills that play a vital role in the process of communication. As listening can be seen as fundamental to all communication, poor listening can become a major barrier to communication. It can result in break-down of communication, or wrong, improper and incomplete communication. Messages can be lost, misunderstandings may crop up, and people may perceive or be perceived wrongly.

HEARING AND LISTENING

Most of the problems discussed above crop up because we do not discern between the two activities, *listening* and *hearing*. Hearing is primarily a physical act that depends on the ears.

Unless there is a physical disability or problems such as noise or distance, it happens automatically. It requires no special effort from the listener. 'Listening', on the other hand, is a much more conscious activity that demands a lot more than just hearing. It requires the conscious involvement of the listener, the acknowledgement of understanding and response. The listener has to hear, analyze, judge, and conclude. When a person is listening, he is constructing a parallel message based on the sound clues and verifying whether his message corresponds with what he hears. Hence, listening is an active process in which the listener plays an active part in constructing the overall message that is eventually exchanged between a listener and a speaker. It is a process that actively engages the speaker as well as the listener. Both are equally involved. Even as the listener is listening, he has to process the facts, study the body language of the speaker and also project the appropriate body language to the speaker. The speaker, in turn, has to cognize the feedback given by the listener and respond. It is, in other words, like a sea-saw, where both the listener and the speaker monitor one another's response and then act. A person, who listens well and engineers his body language appropriately, is seen as a 'good conversationalist' even though he actually speaks less. This is active listening.

> *One of the most valuable things we can do to heal one another is listen to each other's stories.*
>
> *—Rebecca Falls*

ACTIVE LISTENING

Most of the problems in 'listening' arise because of the discrepancy in our speed of talking and listening. On an average, we can speak around 120 to 150 words a minute. But the brain is capable of processing 500 to 750 words a minute. Most of the brain is idle when we are just listening. Attention, thus, gets dissipated and the mind starts getting engaged in other things. As a result, our listening becomes partial and selective. Often instead of listening and trying to understand what the other person is saying, we get more involved in forming our counter arguments. This also becomes a kind of selective listening where, more than listening, we are involved in our own response. In effective and active listening, the listener, after grasping the content of the speaker gets engaged in trying to understand him. He looks at the problem from the other person's perspective, engineers his body language appropriately giving the listener constant feedback. This process, as mentioned earlier, is as engaging as talking. Thus, it leaves no space in the listener's mind empty for speculation or formation of anti-discourse.

ACTIVITIES

1. In a group, find out one person who loves cricket and another who does not. Make them sit facing one another and give the following instructions.

For a couple of minutes, think about the reasons why you like or dislike cricket. Start giving your views simultaneously. Rather than listening to your opponent's viewpoint, try to impress your views on him. Continue doing this for about 10 minutes.

When the task is complete, take a feedback of what went on from the two participants. Take the opinion of the audience too.

2. Make a slight variation of the same exercise. Instead of ignoring the other viewpoint, each participant should carefully listen to what the other is saying, paraphrase it, and then put forth his own point. The conversation of each, thus, should begin with this pattern:

So you feel that ______________ .
but I feel differently. I believe that ______________ .

At the end of the exercise the audience should note the difference in the body language of the listeners in the two exercises. The differences can be analyzed along the following parameters.

	Activity 1	Activity 2
1. Eye contact		
2. Facial expression		
3. Sitting posture		
4. Head movements		

Task for the two participants : Enact the situations. At the end of the performance each of you should note down your feelings about the other. Now read one another's role-description and re-enact the situation. Is there any difference in the way you feel for one another now?

Task for the audience : After watching both the performances, analyze the kind of listening the teacher did in the first performance. Did it become different the second time? Give reasons for the change if there had been any.

Take more topics like classical music, network-marketing or job-hopping, select people with strong opinions for and against the topic and repeat the exercise. In every case, note the difference in the body language during the two exercises. Make a list of the common differences you found in the body-movements. Anyone from the audience can make a brief presentation on it. Include in the list any commonality in the feelings that the participants reported. Put together, these can be seen as the basic differences between how people react during hearing and listening.

> *"One of the best ways to persuade others is with your ears—by listening to them."*
> *—Dean Rusk*

KINDS OF LISTENING

The exercises make it evident that there are different kinds of listening. Depending on the quality of listening, it has been divided into four types.

1. Ignoring
2. Selective listening
3. Attentive listening
4. Empathetic listening

Ignoring: This is the kind of listening where the listener is entirely ignoring the message as well as the message giver. He/she might just be 'pretending' to listen while doing or thinking something else. This can be very damaging because the listener's lack of participation becomes evident through the body language. The speaker might feel snubbed and hurt, which might further lead to a total break-down of communication. The same preoccupation might also result in the listener not taking note of the speaker's reaction.

Selective Listening: Selective listening is listening to parts of the conversation while ignoring most of it. This is the kind of listening we practise often while listening to repeated public announcements or the TV news if we are looking for some specific information. If we are waiting for news about the cancellation of trains in a certain route, for example, extensive coverage about a cricket match or weather is most likely to be at the fringe of hearing. We register the broad topic at times but the details are ignored. The brain registers the topics and then dismisses them or just 'shuts off'. This often happens in classrooms too. Many students practise selective listening. The whole lecture is rarely listened to with the same intensity. Individual students pick up topics of their concern or interest in a lecture and pay close attention to it. The rest of the content is either given peripheral attention or ignored. It is only sometimes that a whole lecture is absorbed similarly.

It is interesting to note the instantaneous change in the body language when the listener moves from the non-listening to the listening phase. The facial expression becomes more focused. The eyes, especially, show a lot more concentration. The listener might even lean forward in the chair or towards the speaker or might straighten up and turn towards the direction the message is coming from. When the message has been observed, the body language relaxes visibly. This can be noticed in the slumping of the shoulders and diminishing eye-contact. During any conversation, it is very important for the speaker, especially, to look out for these signs. If the listener or listeners are listening only selectively, the structuring of the content may need to be altered; the material may have to be made more relevant, or repetitions may have to be avoided. But if you are a listener engaged in a conversation with a speaker, beware. Your body language will most probably give away that you are listening only partially!

Attentive Listening: Attentive listening in a kind of listening where there is no selective dismissal. The listener listens to the speaker completely, attentively, without glossing over or ignoring any part of the speech. This is the kind of listening we find when there is a discussion, for example, on a topic we are interested in or we are critically examining a piece of information for further discussion. Critical listening allows us to form an opinion on the topic being discussed and even design our response appropriately. It allows us to assess the viewpoint, the perspective of the speaker and weigh the arguments appropriately.

Empathetic Listening: This is the ultimate kind of listening that is done not just to listen and understand, but understand the speaker's world as he sees it. Here, one empathizes with the speaker, understands his viewpoint but does not necessarily agree with him. This kind of listening has almost a therapeutic effect on both the speaker and the listener.

Empathetic listening is different from attentive listening or critical listening. As Stephen Covey puts it, here listening gets into "another person's frame of reference. It is listening, not only with one's ears, but one's heart". To quote Covey again, "you listen for feeling, for meaning. You use both your right brain as well as your left. You sense, you intuit, you feel".

This is the kind of listening one friend gives to another friend when the latter feels the need to speak, or a sympathetic parent gives to the growing child if he/she has come back from school, troubled.

> *When you are listening to somebody, completely, attentively, then you are listening not only to the words, but also to the feeling of what is being conveyed, to the whole of it, not part of it.*
>
> *—Jiddu Krishnamurti*

Activities

3. Look around you and make a note of situations when you and others listen. You'll find them in plenty! Try and categorize them. Note five situations when you think you and others do the following kinds of listening.

    1. *Critical listening*
        a.
        b.
        c.
        d.
        e.
    2. *Selective listening*
        a.
        b.
        c.
        d.
        e.
    3. *Ignoring*
        a.
        b.
        c.
        d.
        e.

4. Make a note of a situation when you thought that you were adequately listened to. Try and recollect how you felt, using three adjectives to describe your feelings. Also try to recollect what it was in the person that made you feel so. Use three adjectives to describe the attributes of the listener.

    Your feelings
    a.
    b.
    c.

    The listener's behaviour
    a.
    b.
    c.

Recollect five points about the body language and the facial expression of the listener.

a.

b.

c.

d.

e.

Role-Play

The Situation: This is a role-play that shows the conversation between a teacher and a fourth-year engineering student. The student was supposed to have submitted his assignment three days ago. But he has been missing. He has come with it now. (With this introduction, give the role descriptions to each separately. Do not let one look into the description of the other. But in the absence of both, the audience should hear both the role descriptions).

The Teacher: You are a strict disciplinarian. You believe that a deadline is a deadline. Students coming up with reasons for delaying work is a source of constant irritation for you. This time you have decided that you'll not accept assignments if there is any delay. This student coming now can do well if he applies himself. You appreciate him. But often, he disappears from college without any information. You are determined to teach him a lesson. You believe that it is for his good that you are doing this.

The Student: You are a good student but you know that you are under-performing. Your father is an alcoholic and sometimes does not come home for days. At such times you have even gone searching for him. You have not shared this problem with anyone. This time you wanted to finish the assignment on time, you also put in your best. But, again, your father disappeared and you had to go in search of him. You have come back with the assignment as soon as you could. You are tired, irritated, and you now dread meeting this teacher who, you feel, unnecessarily makes things difficult.

> *The people of the world are islands shouting at each other across a sea of misunderstanding.*
>
> —*George Eliot*

BARRIERS TO GOOD LISTENING

Sometime or the other, all of us listen partially, we do selective listening. Some times we also ignore some messages. There are many reasons as to why we do not listen completely. Some of them are the following:

1. Physical reasons : One chief cause of bad listening could be a person's inability to hear properly. Apart from this, noise and distance too could become barriers to listening properly or

not listening at all. Anyone who has tried talking on a running train or pass on a message to someone across a crowded street has experienced what a physical barrier could be.

2. **Age and Attitude:** Age and attitude are sometimes reasons for not listening well. A four-year-old child's constant conversation is likely to be ignored by most parents. A teenaged son or daughter is similarly likely to ignore the parent's constant caution about driving rules, etc. Difference of age often makes one feel that the person speaking cannot possibly have anything interesting or relevant to say. This, often, creates an attitudinal block, which results in the listener ignoring the message or assimilating it partially.

3. **Mental Set:** Sometimes, the listener is already conditioned to think that the speaker will adopt a particular attitude or a line of argument. If a conversation begins with this kind of mind-set, it is obvious that no 'listening' or communication will take place. The listener might entirely ignore what the speaker says or listen to only what he/she thinks the speaker will say. Meanings here will be wrongly inferred and vital parts of the conversation will be skipped.

This kind of a mind-set can be extremely harmful in both professional and personal interaction. If one comes to the negotiation table, for instance, with a closed mind determined to reject the opponent's proposal, there is little chance for the talks to go forward and reach a resolution. In inter-personal situations, similarly, if one is pre-determined to look at a person or his talk in a particular light, there is little chance of our forming a correct opinion about him and his views. Such conditioning often prevents a bad situation from getting better. It makes one blind to the fact that people might be willing to change, or be more accommodative.

4. **Language:** Language can be yet another reason why people do not hear correctly. It could be the problem of a French speaker conversing in English or a Tamilian trying to speak in Hindi. The mother-tongue interference plays a major role and prevents the listener from listening correctly. It is important, therefore, to make sure that we speak the language we are conversing in with reasonable clarity.

While speaking English, especially, it is important to be aware of our pronunciation, tone, pitch, modulation and stress.

Language can sometimes be very context specific. A group of college boys and girls talking in the college canteen, for example, can have an altogether different register. Slang might be used in specific ways and words too might have different codes and meanings. Listening here will mean being familiar with the particular register. Unfamiliarity can become a barrier to listening. In specific knowledge areas and professions certain words have specific meanings. Unless specified, these too can become barriers to listening comprehension. The same is true of in-house acronyms.

5. **Careless Listening:** It is a common sight to see people looking at papers, sifting through lists or even fidgeting with objects like paper weights while listening. This can put the speaker in a very awkward position. He has no clue of what the reaction of the listener is or even whether the listener is listening to him or not. Such actions can be annoying for the speaker. It can also be seen as a way of snubbing or dismissing what the speaker is saying. Often, it can also indicate to the speaker that what he is saying is not important for the listener. This kind of gesture can seriously hamper communication if used by superiors at the workplace or in any inter-personal communication. If the speaker does not feel 'listened to', the act of communication

will always remain incomplete. Listening in such cases, is bound to be partial. Even if the facts are conveyed, understanding of the facts is generally inadequate or incomplete.

Such habits are commonly observed during telephonic conversations. Since the listener is not present right in front, speakers often tend to do paper work, fidget or draw diagrams. The speaker, in fact, should be more careful during a telephonic conversation. The listener has no inputs from the speaker except the voice, the pitch, the modulation and the pauses. Body language and facial expressions are absent in this form of communication. So the language being used, the pitch and modulation, and especially the pauses have to be used very carefully to convey the right shade of communication or even avoid mis-communicating!

Role-Play

This is a situation where an employee has come back from the boss' room agitated and angry. The report, which was to be submitted the previous day is not yet ready. The boss blames the employee for the delay but she feels that her share of work was done long back. Back in the common room, she is talking about it to her colleague and friend. This is a conversation between the friends.

Role Descriptions

Friend–1: You are new to the job but you feel that right from the beginning, your boss has been trying to find faults with your work. Even now you feel that he did not hear what you had to tell him about the work distribution. You are angry and you want to complain to the person next in hierarchy.

Friend–2: You are a good friend of this employee. You have worked longer in the organization and you know your boss as a fairly reasonable and balanced person. You feel there has been some mis-understanding somewhere. You also know that your friend is impulsive. You do not want her to take the matter to the next in authority.

After the role-play has been enacted, have a general discussion on the following points.

1. The communication barriers that you could observe.
2. What caused the barriers?
3. How could they have been avoided?

ACTIVITIES

5. Given below are certain statements and also the responses. Which do you think is the best possible response?

1. An old woman sitting next to you in the train starts complaining about rheumatic pains. "These knees are getting worse. God knows how many more days I'll have to suffer them. Living with such pain is not worth living."

You reply saying:

a. This is common at this age.
b. May be, you haven't seen a good doctor?
c. It must be really painful!

2. An old friend speaks to you about her broken engagement: "I feel betrayed. He just took me for granted. I can't imagine he could be so rude to me. What does he think of himself after all?"

You reply saying:

a. Just forget him. There will be many more ready to wait on you.
b. You're stupid to have believed him.
c. Couldn't you find anyone else.
d. Oh dear! You must be shattered!

3. A small child walks into the room saying, "The clouds outside are strange! Yesterday I saw an elephant today, it is like a mother and a child. There is a flower also!

You say:

a. You are always dreaming.
b. Clouds keep moving, my child. The shapes are just formed to change.
c. Shall I tell you how clouds are formed?
d. It is fantastic! You know, I had once seen the sea and a jumping whale in the cloud.

4. You come back home late after a hard day's work and your husband greets you saying: "It was a terrible day for me. My bike had a flat tyre. I forgot my papers at home. The work had to be redone in office. And when I came back, I found that the kid had not eaten in the afternoon. I've been trying to feed her since then. But she refuses to touch food. She's just stubborn and unreasonable".

You reply saying:

a. I've had my share of problems too today.
b. This girl needs a real spanking!
c. You've been really stressed out today.
d. Can't you see I'm just back?

In each of these cases, analyze what each of the answers would mean and decide which would be the best possible response.

CHINESE WHISPER

Choose five volunteers from the class. Ask four of them to be outside. Read the contents given below to the only volunteer present inside and the class. Ask the people outside to come in one by one. The first person should repeat it (whisper it) to the second, the second to the third and so on. Observe the way the message changes. After the fifth person listens to it, ask him to repeat it once to all present. Compare it with what was said first. (This is done to show how message gets lost while traveling; what we listen to and what we ignore; the manner in which we summarize, interpret, and recreate while listening)

"A scooter was coming at great speed from the south end of the factory and trying to move towards the north-west. Even as it was trying to enter the lane to the left, a truck coming from inside the lane blocked its way. The scooterist tried to overtake but was again stopped by a car coming behind the truck. He came very near to dashing the car. The car driver, thoroughly disgusted with the traffic, came out and cursed the scooterist. Upset with all this, the scooterist turned back with great difficulty and took the next lane."

> *Wisdom is the reward you get for a lifetime of listening when you'd have preferred to talk.*
>
> *—Doug Larson.*

GOOD LISTENING

A good listener will:

1. Try to understand the speaker's perspective. It is not necessary to agree with the speaker, but a good listener will always try to see from the speaker's perspective.
2. Listen with the whole body – As we have seen, the listener is as active a participant in the act of conversation as the speaker is. For the speaker, the body language of the listener is one of the most important sources of getting feedback. The sitting posture, the facial expression and eye contact are important clues for the speaker to go on speaking or to stop. They can encourage, discourage or even snub the speaker. If you want the speaker to feel reassured, listen with your whole body, let the speaker know that you are listening and you understand.
3. Do not judge prematurely – Since the brain can process speech much faster than one can speak, it is easy to think ahead, judge the talk and even evaluate the speaker and his talk. A good listener, however, will always try to look at the speaker's perspective, try to understand why the speaker feels the way he/she feels. If you want to be a good listener, therefore, avoid judging the speaker's talk or personality prematurely. Give some time. Try to understand and then arrive at a conclusion.
4. Go beyond the words of the speaker – As said before, a good listener will always try to understand the speaker's perspective. But more than the words, it is important to understand the spirit, the sentiment that keeps the conversation going. For good listening, thus, it is necessary to not get stuck in the web of words. One has to analyze the context, study the body language, judge the attitude of the speaker and the reasons why one is responding the way he/she is. It is necessary, therefore to go beyond the words of the speaker
5. Paraphrase the speaker – A good speaker, while listening, might also paraphrase the speech of the speaker. This may not be a detailed paraphrasing, but responding in a few words. Adding nothing, changing nothing, asking no questions, just summarizing the speaker's thought and giving information about what has been understood.

Activities

6. Read out the following passage to the class loudly and clearly, and complete the activities given below.

For centuries, the principal raw materials for making paper were cotton and linen fibres obtained from rags. Today, these have been largely replaced by wood pulp. Trees are cut down, taken to the sawmill and chopped up into small pieces. The pieces of wood are then ground up and mixed with water to make wood pulp. The wood pulp is then washed to clean impurities. It may then be bleached to make the paper white. In the next step, starch or clay may be added to improve the surface of the paper. The pulp is then fed into the paper making machine. In this machine, the pulp is spread across large areas of wire mesh- sheets of metal with a large number of holes in them. Here, the water is drained off. The sheet is then passed between rollers, which squeeze the water out.The dry sheet then travels through a series of heated drums.At this stage, a coating may be applied to make the paper smooth and shiny. This process produces a continuous sheet of paper, which is wound into giant rolls.It is then trimmed to remove the rough edges, and cut to the desired width.

1. **Listen to the description of how paper is made and complete the flow chart.**

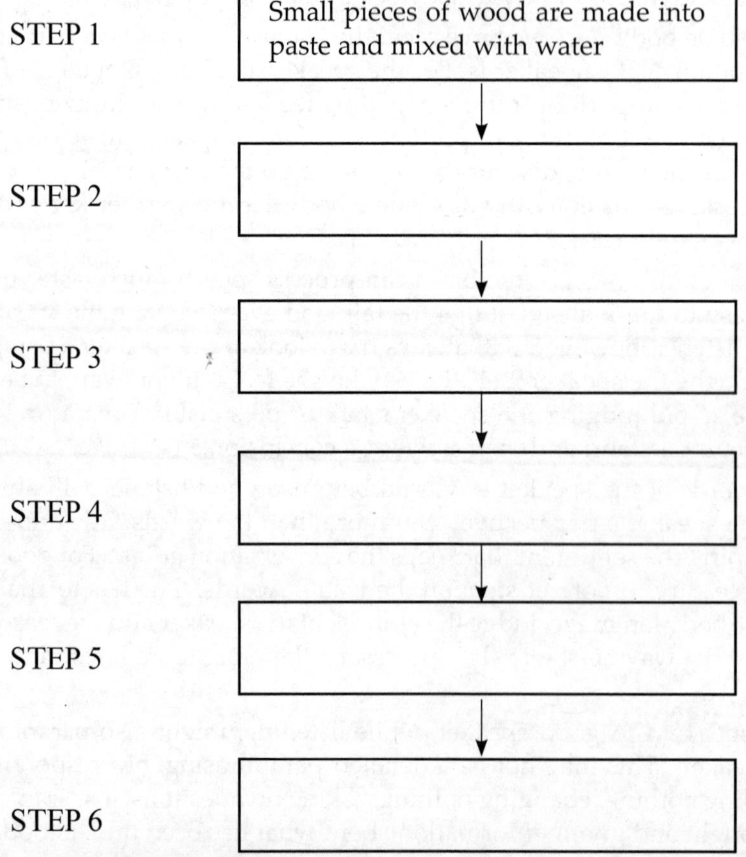

2. **State if the statements given below are true**:
    a. Good pulp is bleached to clean out the impurities.
    b. The paper-making machine has a large wire-mesh on which the pulp is spread.
    c. The pulp is adjusted, trimmed and cut at this stage to remove rough edges.
    d. The sheets are passed between rollers to squeeze out excess water.
    e. The heated drums are primarily used to give a glossy coating to the paper.

7. If you listen to the English of any good speaker carefully, you'll realize that only a part of a word or one part of the sentence is stressed at a time. Proper stress is very important in determining the meaning of a sentence. Read the sentences given below, stressing only one word at a time. Every time you read the sentence, stress only one of the words underlined and discuss the different meanings that emerge.

ST-1: Did Sita tell you that she was leaving?
ST-2: It is surprising that you mentioned this fact.
ST-3: Could you please leave me alone?
ST-4: I've never seen you so angry before.
ST-5: Did Saurav buy that black car?

Build dialogues around the different meanings you find in every sentence each time you stress a different word (Maybe, you could divide the work between five groups and enact them in front of the others)

Here is an example:

ST-1: **Did Sita tell you that she was leaving?**
Stress—**Sita**

A Hey! I just heard that Sita's visa is through.
B Really? But I always thought she was strongly against people leaving the country.
A All that makes good talk, you see. When the opportunity comes who would like to refuse?
B I still have my doubts about the whole story. Did Sita tell you that she was leaving?
(There should be two more dialogues from this sentence, one stressing on "you" and the other on "she". Take into account the changed meaning every time you change the stress and compose the situational dialogues accordingly.)

SUMMARY

- Hearing is distinct from listening. It is listening alone that makes a communication meaningful.
- Listening to somebody with the whole of your attention says, "you matter to me".
- We listen much faster than we can speak. 'Understanding' will fill the gap in speed.
- In listening, the listener is as much involved as the speaker is.
- Careless listening can snub the speaker and halt the entire communication process.

- To listen well, one does not have to agree with the speaker but try and understand the speaker's perspective.
- Since listening can be seen as fundamental to all communication, poor listening can become a major barrier to communication.
- Listening is a much more conscious activity that demands a lot more than just physical hearing.
- A person who listens well and engineers his body language appropriately is considered a good conversationalist even though he actually speaks less.
- An effective and active listener, after grasping the content of the speaker, gets engaged in trying to understand him and looks at the problem from the other person's perspective and engineers his body language appropriately giving the listener constant feedback.
- Selective listening is listening to parts of the conversation while ignoring most of it.
- Attentive listening involves listening to the speaker completely, attentively, without glossing over or ignoring any part of the speech.
- Empathetic listening is the ultimate kind of listening that is done not just to listen and understand, but understand the speaker's world as he sees it. It is getting into another person's frame of reference.
- Most of the problems in listening arise because of the discrepancy in our speed of talking and listening.
- Physical reasons, age and attitude, mental set, language and quality of listening are some of the other factors that become either barriers to or enablers of good listening.

REVIEW QUESTIONS

1. How can we define listening?
2. What is the difference between hearing and listening?
3. Discuss the concept of active listening.
4. What are the four types of listening?
5. What is the difference between ignoring and selective listening?
6. How do you relate attentive listening to empathetic listening?
7. What are the barriers to good listening?
8. Discuss the role of attitude in listening?
9. What do you understand by careless listening?
10. Discuss the role of language in effective listening.

CHAPTER 4

Speaking and Negotiation Skills

> *Three things matter in a speech: who says it, how he says it, and what he says- and, of the three, the last matters the least.*
>
> —*John Morley*

Chapter Objectives

This chapter aims to present a general overview of the speaking skill. Briefly, it dwells on the general factors involved in speaking, and intends to familiarize the reader with the crucial elements that need to be given attention. The chapter discusses the important speech styles, presentation skills, group discussions, role-plays and negotiation. Signposting, attention curve, audiovisual aids, power point presentations and important facts and styles used in communication situations that require negotiation are some of the micro areas that are dealt in this chapter.

THE ART OF SPEAKING

Today's world is one where speech abounds. The electronic media has brought in speech in every form into our houses and our daily lives. In spite of this abundance of talk, however, speaking skill is still a difficult skill for many Indian students. This is because in India, opportunities for practicing speaking in English in an authentic, communicative setting are not sufficient. This is true in spite of the fact that English is widely used in education, career making, business contexts and is also the associate official language of the country.

Opportunities for speaking practice should be provided. Teachers and trainers need to design activities involving pair work, group work, and role-plays and stimulate the students' intrinsic problem-solving abilities. Speaking is an active process and students learning it need to take an active role in developing themselves. Practice will become more meaningful if topics and events from real world are used as starting points for establishing genuine communication among the learners.

SPEECH STYLES

Speaking manifests in society in various styles. Some of these are the formal, informal, polite, normal, strong, blunt, tentative and direct styles.

These styles are context based and it is important for speakers to strike the right attitude and choose the right language. As with many other languages, in English also you have different ways of expressing the same content and message. The style you choose will depend upon some or all of the following:

- The relationship you have with the people you are talking to; (e.g. whether they are close friends, strangers, people in authority, etc.)
- The situation you are in; (at a friend's party, an official reception, etc.)
- The mood you are in; (angry, happy, nervous, etc.)
- The mood of the people you are talking to; (you will probably be careful when talking to a friend who is in a bad mood.)
- What you are talking about; (you will be more careful in your choice of words if you want to complain to a friend about his behaviour than you would if you were inviting him for dinner.)

It is important to choose appropriate ways of saying things according to the situation you are in. In many situations it will be appropriate to use normal or neutral language. In other situations, it is necessary to use language items that are appropriate to special situations.

- We use ***tentative*** language when we are sincerely unsure of our facts or of how we feel. E.g. It's very kind of you to invite me, but I'm not sure if I can come.

We also use ***tentative*** language when we want to give the impression of being unsure in order to be tactful and diplomatic. For example, if we want to disagree with a superior, it would probably be too strong to say "***I can't agree with you***" and it would be more appropriate to be ***tentative*** and say ***I'm not sure if I'd agree with you***."

- ***Direct language*** is the opposite of *tentative* language; it gives the impression that the speaker is very sure. This impression is appropriate if, for example, we want to agree with someone. But it can sound rude and inappropriate in many situations like inviting a superior to a party.

  E.g. i. *No .You are mistaken the statistics just cannot be this high.*

  ii. *I see your point. But we'll have to go ahead with our previous plan.*

- We use ***polite language*** when we want to sound particularly polite without being tentative.

  E.g. i. *I'm sorry. But I feel there is a mistake somewhere. According to my calculations, the statistics should not be so high.*

  ii. *You definitely have a point. But I'm afraid this time we'll have to go ahead. Next time onwards, we'll definitely consider these points.*

- ***Formal language*** creates the impression of social distance between people. It occurs mostly in official situations e.g. business meetings, official receptions.

  E.g. i. *I'd be delighted if you could make it to the party. We'll all look forward to it.*
  ii. *Forgive me if I sound curious. But isn't this the same girl we saw him with yesterday?*

- ***Informal language*** is used basically between friends. It is generally inappropriate to use it with anyone else.

  E.g. i. *Cut it out will you? I've had enough of this.*
  ii. *You 're coming to the party tonight, aren't you? I just wont take 'no' for an answer.*

- ***Strong language*** carries with it a strong sense of conviction. It usually sounds very direct.

  E.g. i. *This is impossible! Hoe could you ever promise without consulting me?*
  ii. *I'm gone without this project. I must get it come what may.*

- ***Blunt language*** is extremely frank. It should be used with extreme care, as in most cases it will simply sound rude.

  E.g. i. *I know you're lying. You can't fool me!*
  ii. *I must tell you. Your work was not up to the mark and we'll have to review your extension.*

In most cases we use normal and neutral language but sometimes, depending on the situations we are in and also on the basis of our co-speakers, we use special language. The type of language we use shows our attitude.

> *Silence, along with modesty, is a great aid to conversation.*
>
> *—Montaigne*

Activities

1. Carefully look at the box given below. It is an illustration of how different language patterns can be used to express the same content.

Purpose: Asking Someone To Do Something.
Could/would you … please?(Polite)
Do you think you could…? (Polite)
I wonder if you could possibly… (Tentative)
Would you mind…? (Formal)
Can/will you…please? (Direct)
Do me a favor and…, will you? (Direct/informal)

2. Match the language snippets given in column A with the appropriate situations given in column B.

	A	B
1	...well, in any case, it was a superb concert.	the person is making a negative comment on the book.
2	In other words, I feel there's too little information in this guide.	somebody is praising a music program which he had attended.
3	there's a lot of rubbish being shown in this serial.	a viewer is commenting on a TV program.
4	Please Rao. I know it's a bit of an unpleasant job. But we have to do it.	a person discussing a very personal problem with a friend
5	I really don't know what to do Raj; I think I'll say no to him.	a person persuading his colleague to take up some task

3. In the following situations, you must decide on the appropriate attitude you would like to convey and the language that you would actually use.

   a. You are on a train and you want the window opened. You would like to ask an elderly gentleman sitting near the window to open it, but you are not sure about his response.

      Suggested answer: you may use tentative language. E.g. I wonder if you could possibly open the window.

   b. You want your water bottle, which is on a chair near your friend. You ask him to give it to you.

   c. A friend of yours has just called up to say that she would come tomorrow evening to meet you. However, being busy yourself, you ask her to come the day after tomorrow.

   d. You are the secretary to the manager of a company. A person has just called up to speak to the manager, but he is in a meeting. You ask the person to call back after an hour.

   e. You have decided to break your engagement with a boy. You parents had decided the match for you and forced you to go through the engagement. Now you have heard certain things about him and you are determined to break away.You want to do it decently without creating bad feelings.

   f. You are tired of instructing your junior about a certain procedure .You feel she's not willing to learn. You want to make it clear that either she has to learn or quit.

   g. Both you and your wife are software professionals. You come home late and work even over weekends. You want to suggest that your wife could quit the job and take up part time assignments.

PRESENTATION SKILLS

> *A speech is poetry: cadence, rhythm, imagery, sweep!*
> *A speech reminds us that words, like children have*
> *power to make dance the dullest beanbag of a heart.*
>
> —*Peggy Noonan*

Effective presentation skills form a very important aspect of our communication skills. It can make or mar a communication situation. For an effective presentation, one has to choose a style that suits one's personality and fits the situation; carefully select and prepare the presentation material; speak effectively and convey the right message.

Making an Effective Presentation

For making an effective presentation, one has to bear the following points in mind:

1. Break down the material into main points. Do not have more than five or six main ideas, as it is generally difficult to remember more at a stretch.
2. Prepare notes or outlines.
3. Arrange them in a logical, coherent manner.
4. Make your information or facts effective. You may include statistics, illustrations, and graphs etc if required.
5. Deduce recommendations based on your main points.
6. Use appropriate visual aids.
7. Deal with possible questions from the audience.

A presentation style can either be formal or informal. During a formal presentation, one has to take care of the following:

Dressing appropriately: The first impression one makes on one's audience is always very vital. The things that you should keep in mind are i) research your audience ii)dress appropriately. Organizations have their own dress code and people generally like people who look like them. So adapt your outfit and be in tune with your audience; iv) Never let your appearance overpower your message. Remember that the audience has come to hear what you have to say, not see you.

Judging the audience: During a presentation, one has to carefully judge the audience- their level of knowledge; the general aptitude etc. Finding out as much as possible about the persons you are going to address will help you to pitch your talk at the right level. This information can be about the audience's age, profession, specialization etc. These have to be the determining factors in designing the talk and deciding the style. Here, one has to also consider whether it is a large or a small audience. A presentation that is suitable for a group of five or six, for example, has to be different from one meant for a group of 20 or 30.

Using the right style and language: A presentation definitely cannot be informal and casual. But at the same time, it cannot be put in extremely formal, stiff or frozen language. Stumbling for words and speaking haltingly will bore an audience. You need to develop fluency and a good command over language and also learn to use it for your maximum advantage. The audience would like to listen to a speaker who has a confident delivery and is in control of the situation. Developing a positive attitude to speaking will go a long way in increasing the confidence levels. The "you approach" too can be helpful. The audience feels a sense of involvement if directly referred to.

Word order: Choose your word order in a way that you present the relevant information clearly and distinctly. Instead of saying, "Britishers, Moghuls, Afghans and Aryans were all invaders in India" it is better to say, "Britishers were invaders, so were the Moghuls, the Afghans and the Aryans."

Keep in mind the fact that the most important information comes in the front or near the front.

Signposting: Sometimes, you know where you are proceeding and how. But your audience is ignorant of it. It is important, thus, to have the important signposts, suggest the direction in which you are moving. This will also help you plant the facts, clearly signify and categorize the message that is to come. The following would be some of the signposts the audience find useful. They are suggested to give the required flow to your presentation:

Expressions often used:

Openings:

I'm here today to report … and to present …
My purpose today is…
My main aim this morning is …
The title of my presentation is …
My topic today is…

Organization:

I have divided my presentation into _____ sections.
The first point I wish to make is …
The first part of my presentation will deal with …
Firstly …
Secondly, I want to …
Finally I'd like to talk about …
In conclusion …
To conclude …

Drawing attention:

As you will notice, …
As you can see, …
I'd like to draw your attention to….
You will note that….

Linking:

As I said earlier, …
As I mentioned earlier, …
We shall move on to….
Later I'll be talking about …
In addition to …
For this reason …
On the other hand …
To look at some of the other options …
Some advantages and disadvantages in the proposition are …
If this is one side of the picture, the other side is …
If we look at the picture holistically we find …

Changing subject:

Moving on to the question of …
Let me now turn to …

Emphasizing:

What we have to realize is ….
What I find most interesting is ….
What I'm getting at is …

Introducing evidence:

If you look at …
Let me show you that …
Let me explain …

Making recommendations:

I strongly urge that …
You ought to …
I recommend that …
I think you should …

Summing up:

Let me now sum up.
To summarize my main points…
To recapitulate

Points to Remember

1. Communicate at the right level with people
2. Select the right style for the occasion. Decide whether the situation warrants a formal or informal style.
3. Moderate your speed.
4. Use your body language to make people feel at ease
5. Be in control of the situation

The Content of Presentations

During a presentation it is also important to prepare the presentation material with care. The important points here are:

1. **Researching the subject**: It is important here to be clear about the objectives of the presentation, about the audience you are presenting to.
2. **Selecting the content**: Once the information has been gathered, it is necessary to filter out the essential points. One has to group the ideas under separate headings; classify the information depending on the available time; keep the matter strictly to the point.
3. **Planning for the talk**: To get the message effectively across, one has to carefully draw out a presentation layout. A well-planned presentation is always a well-received one. The important factors to be considered at this stage are the following:
    a. **The Beginning**: - During a presentation one is always sure of the first few minutes of the audience's attention. One therefore has to be very careful about the beginning; make an impression that will hold the attention of people. One can always start with a quotation, a question, a dialogue or even an anecdote, a fable or a parable. A joke, an unusual definition or a startling statement or statistics too can be effective at the beginning.
    b. **The Middle**: After making an impressive beginning one has to be able to deliver the contents effectively. The contents should be well structured, they should be logically connected and effectively lead towards a specific goal. To sustain the interest of the audience, it is important here to include examples and personal experiences, which will make the material authentic and interesting.
    c. **The End**: The way a presentation ends is again vary important. Primarily, this is what the audience will remember the presentation as. It is important therefore to give a presentation the right emphatic conclusion that will make a lasting impact on the listeners.

Attention Curve

The attention of the audience during a presentation generally goes through an attention curve. It starts on a high, drops a little first and more steeply later. It rises again towards the end and further up for the last few minutes. Some of the ways the audience can be kept interested are the following:

1. Look into points where the attention curve drops and consider ways of varying the texture. (If your presentation has been largely speech, bring in an audio-visual slide, or have an interactive session that will ensure participation).
2. Keep the sections short, and ensure that every section ends on a high.
3. Get the audience involved decide what you want them to remember and stress on it.

Factors Aiding Effective Presentation

1. **Use of audio-visual aids**: During a presentation, audio-visual aids help ensure the attention of the listeners. It creates the necessary shift of attention and even increases interaction between the presenter and the audience.
2. **Eye Contact**: Eye contact forms one of the most essential means of maintaining a rapport with the audience, receiving feedback and holding attention.
3. **Intonation**: The presenter has to use the right intonation with proper emphasis and stress at the right points to convey the spirit of the message to be delivered.
4. **Body Movements**: During a presentation, the body movements too have to be carefully monitored. They should neither show lack of confidence nor be aggressive. Assertive attitude with the right facial expression and posture is ideal.
5. **The Space**: Depending on the situation and context of presentation the presenter has to constantly negotiate the space between her/him and the audience. Preferably, he/she should avoid the "public space", and use more of the "social space". Rather than being static at one spot the presenter should be able to move and negotiate the points as well as the space with ease.
6. **Words/Phrasing**: The choice of words and phrasing can be very important in a presentation. Some of the general principles are:
   a. Don't use abstract or vague words.
   b. Use active rather than passive sentences.
   c. Cut out jargons or cliched phrases.
   d. Adopt the "you" approach. Wherever possible, replace the third person with the second person.
   e. Bring in personal examples and experiences wherever you can.
   f. Keep the main points as near the beginning of the sentence as possible.
   g. Talk of the way you position yourself and indicate with linkers the way you are moving.

> *How is it that our memory is good enough to retain the least triviality that happens to us, and not good enough to recollect how often we've told it to the same person.*
>
> —*La Rochefoucauld*

Activities

4. Pick out five public speakers whom you often view on the TV. Can you isolate features that distinguish them from one another? Try to find points of commonality and features that make them endearing.
5. Every movement or gesture of ours is a presentation of ourselves, either consciously or sub-consciously—Discuss.

Use of Visual Aids

> *A room hung with pictures is a room hung with thoughts.*
> *—Joshua Reynolds*

You must have all noticed that the use of pictures or visual aids always helps memory. It hel in retaining one's attention and also promotes understanding. In a presentation, similarly, tl use of visual aids makes a better effect on the audience. They help us remember the conten better and assimilate the matter more effectively. Sometimes visual aids save presentatic time and also make the presenter's work easier. Some of the commonly used visual aids are

- Flip charts
- Overhead projector
- Slides.
- PowerPoint presentations

Flip Charts

Flip charts are blank or prepared sheets of charts that are put up during a presentation visual aids. If there is a diagram or sequence of pictures to be projected, you can have pre- pr pared flip charts. Sometimes flip charts can also be used if you need audience participation can be used to generate ideas or quickly record the responses that can be later organized ar ordered. Some of the guidelines while using the flip chart are:

- Diagrams on the flip chart have to be clear and attractive. Colors are more effecti when used on white rather than colored paper.
- Matter on the flip chart has to be visible to all the participants or audience. So spaci between diagrams should be carefully arranged.
- If possible, do not crowd more than one diagram in a single chart. Have separate cha for each diagram.
- If there are many charts, carry an appropriate box where all the charts can be kept.
- Always keep a spare pen in your pocket.

Overhead Projector (OHP)

The OHP like the flip chart can be used in two ways. It can either be prepared in advance written during the presentation as an alternative to using the flipchart. Like the writing in t flip chart, the transparency sheet too has to be carefully prepared. Some of the factors to kept in mind while making an OHP presentation are:

- Use transparencies to show only the important points
- Do not crowd the transparency sheet with too many points on a single sheet.
- Do not switch on the light until the slide is in position, and switch it off before y move it away.
- Line up the OHP before you start the presentation and keep it in the right position.
- Control the lighting effectively. Make sure that it is not too bright or dim.

- Rehearse thoroughly so that you get used to placing and removing the slides without any awkward movement.
- Take care not to stand between the projected picture and the audience.
- Take care to see that you use the right colors that are bright and aid clarity.
- Organize the information you have to give under main heads and subheads. Remember that the organization you give will help organize the matter in the participant's mind.

PowerPoint Presentations

PowerPoint presentations today are the most frequently used for presentations. Power points operate through slides and they have to be prepared very carefully. Sometimes they can make or mar a presentation. They lessen the flexibility of a presentation. But the impact they have can create a lot of interest; bring in variety and life into a presentation. If they go wrong however, the effect can be anything between mild confusion to total catastrophe. So they have to be used very carefully with a lot of planning and expertise. Some of the factors one has to keep in mind while doing PowerPoint slides are:

- Ensure that the slides are an integral part of the presentation.
- Space them intelligently so that they come in at regular intervals but more frequently towards the end. This will keep the interest alive.
- Think and plan of the most important point you want your audience to take back with them. Identify the slide or slides you must include and ensure that they are striking enough to remain in the mind
- Missing an important slide is as dangerous as including unnecessary ones. So plan out every slide carefully and thoughtfully.
- Never have a verbal slide that will have whole statements, sometimes several of them numbered sequentially. This can spoil the readability.
- Do not put into a slide what you will say. All people listen at more or less the same speed but read at different speeds. To keep the audience together, thus it is necessary that they hear things together rather than read.
- Follow a simple principle. Make your writing big enough and obvious.
- Keep the matter- Words, charts and drawings- simple and clear.
- Use the colors very carefully. Always colored backgrounds are not more effective than plain white ones. Also, try to integrate the colors meaningfully. Every color has significance. Understand it before using it.

Points to Remember

- A picture is worth a thousand words.
- Pictures save time.
- They bring in variety.
- They add impact.
- They remain in memory long after words have left

Activities

6. You are the leader of a team of software engineers who have devised an automated system to monitor student attendance and performance. The previous manual system had the following problems:
   - The large numbers made the systematic monitoring of attendance almost impossible.
   - Their marks in internals etc. were being separately monitored and it was difficult to have an idea of student's overall performance.

   Your software has been able to solve these problems by doing the following
   - It can make the daily attendance entry easy.
   - It can show the monthly performance of each student separately in terms of attendance and marks.
   - It can even give a picture of the performance of entire sections, subject wise.

Before installing the system, you have to make an OHP presentation of it in front of the principal and the staff, most of who are not very well versed in computers. Prepare it, make it in front of your friends, and find out from them how effective you have been.

7. Given below is a short write up on Communication. Convert it first into a presentation with transparency sheets for OHP presentation and later into slides for a Power Point presentation. Frame the text and choose the templates and decide on the matter to be put in the transparency sheets. Note how the two mediums of presentation need differential treatment even though the matter is the same:

   "Communication is a multidimensional word that covers everything- interaction with others; casual conversation, persuading, teaching, and negotiating. You cannot communicate with a toy, anything you do is meaningless, it gets no response. When you communicate with another person, you see their response, and react with your own thoughts, feelings and actions. Your ongoing behaviour is determined by your internal responses to what you see and hear. It is only by paying attention to the other person that you get any idea about what to say or do next.

   You communicate with your words, with your voice tonality, and with your postures, gestures, and expressions. You cannot fail to communicate. Some message is conveyed even if you say nothing and keep still. So communication involves a message that passes from one person to another. How do you know that the message you give is the message they receive? You must have had the experience of making a neutral remark to someone, and being amazed at the meanings they read into it.

   Communication is so much more than the words we say. These form only a small part of our expressiveness as human beings, Research shows that in a presentation before a group of people, 55% of the impact is determined by your body language -posture, gestures and eye contact -38% by your tone of voice, and only 7% by the content of your presentation".

   (Adapted from Mehrabian and Ferris, 'Inference of Attitudes from Non-verbal Communication in Two Channels' in *The Journal of Counseling Psychology*)

Role-Plays

Introduction

Role plays are a very effective tool in practicing speaking skills at all levels .At the introductory level they ensure a level of involvement where the participants can start speaking without much hesitation and inhibition. If the characters and situation is well drawn, they also provide the participants a strong support on which they can depend upon and draw material for a good speaking session. At a higher level they can provide good background material to discuss interpersonal skills, body language, pitch, volume, etc Role plays can also be an important tool to teach etiquette in different social contexts.

- Given below are role-plays that you need to enact in groups. Understand the characters and frame a rough speech pattern. Do not frame the dialogues in advance. Respond to the situation as and when it develops. These role plays could be enacted after completing the chapters as indicated.

Role-play - 1

Number of Participants: 4

The Situation: The parents and sister (Saroja) of a 19 year old boy (Sanjiv) do not like the friends he moves around with. They are highly critical of the group and advise the boy to move away from them. The boy however finds no fault with any of them. He appreciates them for what they are. He is further agitated because his friends' families don't bother much about who they move around with. He sees this as an unnecessary interference. This is a scene where all the four (father, mother, brother and sister) are sitting together and discussing the boy's poor results. Naturally, the discussion leads to Sanjiv's friends. The parents are trying to convince him about what is good for him. But he is stubborn and tries his best to convince them that both he and his friends are reasonable. The sister Saroja is the youngest. She is angry with all his friends. She has come to know that they tease her friends in college. She has also seen them smoking in the canteen and knows that teachers are highly critical of their general attitude and behaviour.

Role Descriptions

The Father : You are a 50-year-old senior government employee in a responsible position. You have come up the hard way in life and you want that your children too should make a mark in their lives. You have given them your best and you want them to give their best. Your son's poor results as well as his general careless attitude is disturbing you. You are known for your composure and rational behaviour. But this situation seems to be going out of your control. You are on the verge of losing control over your temper.

The Mother : You are a 47-year old woman working as the principal of a school. You are also an expert in child psychology. You were always proud of the way you brought up your children.

This turn of events suddenly baffles you. You don't want to become authoritative and rule over your son's life. At the same time, you feel that he has to understand his priorities. You're trying the balance the situation at home.

Saroja : You are the youngest child at home. You are 16 years old and recently joined the same college as your brother. You have seen his friends misbehaving and ragging your friends when you joined college. You are very agitated because you have come to know that your brother is friends with the same group. You were deeply concerned and had to come and report about his friends to your parents. You and your brother have always been good friends. Now, he is deeply hurt by your action. You did this after talking to him a couple of times. But he showed no signs of improvement. In this argument, you are caught in between. You want to protect your brother but at the same time you are agitated because he insists on keeping bad company.

Sanjiv : Age: 19. You are irritated and angry with your parents and sister's accusations. Your friends might be a little energetic and fun loving. But there is nothing basically wrong with them. You also feel that Saroja is unnecessarily over reacting.

Role-Play - 2

Number of Participants: 5

The Situation : Situation in a family where there is the father-in-law (Mohan) the mother-n-law (Savitri) the son (Rajiv) and the daughter-in-law (Seema). Unfortunately, the son and the daughter-in-law have no children and cannot have any. The daughter-in-law wants to adopt a child from the orphanage. Her husband too is willing and supportive. The in-laws however are dead against the proposal. They feel that if at all a child is to be adopted it should be from within the family. Preferably, their daughter's child. There is an old aunt (father-in-law's sister, Parvati) also in the house. She agrees with the daughter-in-law in her decision. She has been trying her best to bring about some balance of opinion in the family by negotiating with her brother and brother's wife. She hates to see the daughter-in-law suffer.

This is one day in the family when Seema and Rajiv have identified a child in an orphanage. It is a girl child, less than 6 months old. This is just what they wanted. Seema is happy but apprehensive about her in-laws' reaction. Rajiv feels he can convince his parents. The situation, however, turns unpleasant and both Rajiv and Mohan lose control over their tempers.

Role Descriptions

Mohan : You are 65 years old. You have all your life believed and taken pride in your superior family lineage. You've been the unquestionable head of the family. In fact, your wish has always been a command. You are worried about your son not having a son to carry forward the family line. But you don't understand his desire to "adopt" an outsider, moreover, a 'girl'. You are also disturbed and agitated because you feel that of late your son has started defying you. You're seriously thinking of going back to your ancestral village.

Savitri : You are 57 years old. You've been brought up the old fashioned way. Your husband's wishes are the only "truths" for you. You like your daughter-in-law as a person and even believe in what she wants. But somewhere you dislike that both she and your son are all out to go against the wishes of your husband. Rajiv, being the youngest is your favorite and you feel you can make him come around and listen to his father.

Rajiv : You are a 36 years old, successful and dynamic interior designer, you've traveled widely and moved away from the orthodox values of your family. You love your wife, respect her for her views. You're determined to stand by her in her decision. You've always thought that your father was unreasonable. You strongly believe that you and your wife should have the independence to decide on certain issues. You are the only son of your parents.

Seema : You are 33 years old. You started your career as an Interior Designer. But for some time now you've stopped working because it was creating a lot of controversy at home. You've plans of again starting your career seriously once the dust settles. You were disappointed when the doctor told you and your husband that you couldn't have your own children. But you always believed in adoption and you are happy that you've been able to convince your husband. Your husband always wanted a daughter and you also feel the same way. Since boys are anyway more sought after it is wonderful to give a "home" to a girl child. You don't want to create a rift in the family but the attitude of your parents-in- law is beginning to unsettle you.

Parvati : You are a 58 years old child widow. You've always thought that Rajiv has got his independent and progressive thinking from you. As a young woman you had tried doing many things. You had even started a cooperative society for the woman laborers in your ancestral village. But your brother's high handed, patriarchal views have always come in the way. Your dependence on him has prevented you from rebelling openly. Now you feel Seema and Rajiv should be allowed to live their life. Thanks to Soma's encouragement, you are now associated with many voluntary organizations and you've picked up the courage to disagree with your brother. In a way, you now want to start living your life; move out of your brother's shadow and establish yourself. You are soft spoken by nature. But now you are strong and determined. You have a quiet strength in you.

Role-Play - 3

Number of Participants: 4

The Situation : This is a typical corporate scenario. There is a new H.R. Officer (Arundhati) in office who wants the company to invest in staff training and welfare programs. The company is going into a slump and she feels the training programs will boost the morale of the employees, which, in turn, will have an effect on the production. The head of the finance department, (Ravi) however, is totally against this proposal. He believes in a different work culture which others consider old fashioned.

According to him, this expenditure should be avoided till the situation looks up. He has always had a big say at the decision-making level. But of late he thinks the new M.D. (Sourav) is caught up with new fanciful ideas and his own proposals are being sidelined.

This is a board meeting where the budget for the next half-year is being discussed. Apart from the three people mentioned above, the head of the marketing division (Rajiv) is also present here. The M.D. has brought forth the proposal for new expenditure in training. The new H.R.O is trying to justify the expenditure even at the time of financial crunch. But the head of Finance opposes the move. He has his reasons for it. In the course of the discussion he even gets personal and vents his frustration at being overlooked. (Decide what kind of conversation you want to have. Be careful about the language you choose)

Role Descriptions

Ravi : Age: 47. You have been in this company for the last 10 years and have seen the place grow. In fact, you've had an important role in its growth. Now you are worried about the financial crunch and you are doing all you can to save and manage the finances well. The change over in M.Ds had unsettled you at the beginning but you've pushed it to the back of your mind now. But this new expenditure irks you. You feel it is entirely unnecessary. In fact, the new H.R. Officer too looks unnecessary. You are surprised that the M.D. has decided to take her suggestions seriously in spite of your disapproval. You are concerned about the finances but the whole episode is also a blow to your ego. You've decided to take on the matter headlong.

Arundhati : Age: 26. Youe graduated from one of the premier institutions of the country. You're always been a promising student and your two previous short stints of work have also been satisfying. You are confident that your plan of action can help in pulling the company out of the slump it is going through. The employees have to be re-charged. Their morale has to be boosted to take up any new initiative successfully. Mr. Ravi, however, seems to have come up as a major stumbling block. You've tried to reason out with him a couple of times but he is very convinced about not letting your initiative take off. Personally, you wouldn't like to hurt and insult him. But your work is important too!

Saurav : Age: 34. You have joined the company 6 months ago only to realize what a bad time it has been going through. You were taken up with the reputation of the company before joining. Mr Ravi is a senior employee and you also depend on him heavily. The new recruit Ms Arundhati has some very good ideas, which you'd like to try out. But Mr Ravi is against it and you feel that he's unreasonable. At times he even seems to question your authority and you don't like it.

Rajiv : Age: 35. You are the head of the marketing division and at this time of slump you feel there are more important things to concentrate on and this major rift is uncalled for. You are not against the training program but you don't think it is very important either. You are irked both with Ravi and Arundhati for making matters difficult for everyone. You have some very important matters to discuss but the whole meeting seems stuck to one issue only.

Role-Play – 4

Number of Participants: 4

The Situation : This is a high level confidential meeting of the cabinet committee. The meeting has been called to discuss the effects of a new drug. It is a tasteless, colorless drug that can induce telepathy. The person who takes the drug will experience the emotions of people around him. The only condition for the drug to work is that, there should be physical proximity between the person who takes it and the people whose emotion is to be experienced. As soon as this fact was made known, the scientist who invented the drug has been kept under strict observation and provided with Z category security. There has been mixed reaction in the top government circles too. Some people welcome it; some are strongly against it. All agree that it can be dangerous and has to be used very judiciously. This meeting has been called today to discuss what has to be done next.

Role Descriptions

The Prime Minister : Age: 70.You are quite happy with the drug. You want it to be made available for people occupying top political positions, especially those whose lives are under threat. Basically you are a person who had to employ unfair means to get your post. You know that a lot of people around are against you. And you're eager to know what they are thinking and if they have any plans to de-stabilize you. You are arrogant and impatient. You can tolerate no opposition. Scheming and plotting is your second nature. You're eager to procure the drug and you are also the deciding authority as to who should be allowed to take it.

The Defence Minister : Age : 68. Known for your integrity and values. Rational thinking and anticipating correctly are your strong points. You feel this kind of a drug might spell disaster. Your experience tells you that every person is a bundle of contradictions. There are times when it is better for others not to read your thoughts. It can create unnecessary problems and further damage interpersonal relationships. It would become more difficult to negotiate issues with your opponents because knowledge of their negative emotions can affect your judgment. You strongly feel that the whole story should be put to an end and the drug should never be used. You suspect somewhere that the corrupt PM wants to use it for his own benefit and you are keen to stop him.

The Foreign Minister : Age: 68. You've been a collegemate and also a close friend of the P.M. since childhood. He has helped you at many crucial points in your life. You share memories of your childhood and youth with him. It is a bond you cannot break away from. But you also disapprove of some of his moves. Personally, you wouldn't like him to know that you have mixed feelings for him. And you know that once the drug is made available he'll use it to test everybody, including you. Also, you agree with the defence minister that diplomacy will become impossible once you achieve the ability to access the feelings and emotions of your opponents. For both these reasons, you are eager to stop the entire process of procuring the drug.

The Cultural Affairs Minister : Age: 47. You started your career as an actor. Even after political success, you're still an actor at heart. As an actor, it has been your habit to experience many emotions and act them out. Now this prospect fascinates you. You feel that this capacity to experience others' emotions can make acting easier and more authentic. Also, you feel that knowing and understanding the feelings and emotions of people belonging to different cultures can boost inter-cultural interaction in any country that has diverse cultures. You are keen, therefore, to try out the drug yourself first and then use it on other people subsequently as and when necessary. You've been an actor for most of your life. So you talk rather dramatically, and tend to dramatize your emotions whenever there is an opportunity.

Role-Play - 5

Number of Participants: 3

The Situation : A young lady, who is a computer professional (Karuna) has got an opportunity to take up an important assignment which will involve constant globe trotting for a couple of years. It is a golden opportunity and it can give her all the exposure she needs to excel. It can equip her with the latest know-how worldwide, boost up her C.V. and open up many opportunities in future. Her parents have always encouraged her in all she did. They are happy about this also. But they would like her to get engaged to a boy they have settled for her and then go about her career. The boy's family too is eager to finish the engagement. Karuna, however is unwilling to commit herself in such a hurry. She is not opposed to marriage as such. But she feels that she should be given some time to accept the boy and the idea of marrying him. The boy (Anubhav) is reasonable. He understands both the perspectives. He is willing to wait if Karuna insists. But he doesn't mind getting engaged either. Either way, he has left the final decision to her.

This is a situation where Anubhav and his parents had come to Karuna's house, spoken to Karuna and her parents and left. Very categorically, Anubhav has mentioned before leaving that Karuna need not come under pressure. Nonetheless, his parents have expressed the desire to have a small engagement ceremony before she leaves. Karuna and her parents are discussing the matter now.

Role Descriptions

Karuna's Father : Age: 57. You are a senior executive of a company. Your only dream in life now is to see your daughter happily settled-- not only in marriage but in career too. Anubhav, you feel, is a dream come true. Personally if you wouldn't mind if Karuna got engaged later, but Anubhav's parents and your wife are insistent. You have always been a soft, tender parent and your daughter has always mattered more than anything else to you. Now, you too are in a dilemma.

Karuna Mother: You are 55 years old. You have always been a housewife. Karuna is your only child and you are eager to see her settled in marriage. You are proud because your daughter has had such a flying start in her career. But you feel that marriage is equally important. You are very happy with Anubhav and his family and you just don't want to let this opportunity go. You know that your daughter is frank, sometimes too outspoken. She even has very firm beliefs and convictions. So she will not be happy with just any boy. She needs someone who is understanding, co-operative and broad-minded. You feel Anubhav has all these qualities. Moreover, he seems to really like Karuna a lot. You're happy that he's prepared to wait but you don't want to delay the engagement any longer. You want to feel sure and secure for Karuna.

Karuna : Age: 28. Life has really been playing hide-and-seek with you of late. You were on the ninth cloud when you got this marvelous offer but even before you could start celebrating, this whole question of marriage has started troubling you. You like Anubhav, you respect him as a person but you're not sure if you'd like to marry him. You feel it is unfair to go through the engagement when you yourself are not very convinced. You can explain your position to your father and he'll understand. But your mother is rather adamant on the issue. The last thing you'd like to do is hurt her but you don't like her feeling so insecure on your behalf. After all, you are old enough and independent and why should Anubhav suddenly become a source of security for you? You're in a real fix.

Role-Play - 6

Number of Participants: 5

The Situation : A group of final year students are having a party after the final results are out. All of them are destined to move in different directions. Some of them can meet frequently, some can't. They are all happy and excited abut starting a new life but there is also grief over the parting. In all this, one of them, Anil, has got drunk and started behaving crazy. It's past midnight. All are ready to leave but looking at Anil's situation his friends feel it is not safe to let him drive. They want to drive his car, reach him home and only then go to their own houses. But suddenly, Anil is insistent that he can drive back and take care of himself. This is a situation where two of his friends (Rajesh and Krishna) are trying to tell him that they need a lift and since Rajesh can drive, Anil might as well rest. A couple of others too are trying to help but Anil is adamant.

Role Descriptions

Rajesh : You are 22 years old. You've got admission into a premier institution for the MS program. You have to leave the next day and there is a lot of preparation to do. You're already late and your mother must be waiting for you to begin your packing. You want something to be decided soon so that you can leave. You've always disapproved of Anil's 'drinking habit and even fought with him a couple of times about it.

Krishna : Aged 22, you have always been good friends with Anil. He has some eccentricities but at heart he is a very good person and a true friend. You're very eager to see him safe at home. You don't at all approve of his driving in this state.

Sanjiv : Aged 21, you are part of the group. You've also had drinks but can hold it. You too have to drive back. Anil has got stuck to the argument that if you can drive, he too can. You're trying hard to convince him.

Sanatan : Aged 21, you are a calm person, known for your negotiating abilities. You too have to take a late morning flight the next day and you want to hurry up and leave. You're trying your best to find a way out.

Anil : You have just got a letter that your application for a job has been rejected. Earlier, you also failed to get admission into an institution you wanted to. Your friends have always been sympathetic and helpful. Today in the party, however, you've been overwhelmed with self-pity. Now that your friends feel you can't drive back, you want' to prove a point to them all. It's your way of telling them that you can take care of yourself, whatever be the condition you're in.

Role-Play - 7

Number of Participants: 4

The Situation : A world-class football team is going through a rather bad phase. They had brought home medals in the past. They were known for the team synergy and their excellent performing standards. There was always a very high rate of dependence on one another. Of late, personal problems in the lives of a couple of players have affected their performance. Since they were the key players in the team, the performance, in general, has declined. The team first lost a few times and now, losing has become a habit. The team's morale is very low and the selection committee is under a lot of pressure to change the combination. They too feel that changing some of the members will be for the good. But these are the key members and removing them might give rise to public uproar. The coach of the team feels that removing players will effect the team performance in a big way. Instead, it would help if people are given some time to recover, set things right at the personal front.

This is a discussion of the Selection Board, the Minister of Sports and Cultural Affairs and the Coach.

Role Descriptions

The Minister : Age 50. You have a point to prove to your predecessor and for that the team has to perform and win medals for the country. Winning is all that matters to you. You are ruthless and believe in results. You are even out to change the selection committee if they hesitate to take hard decisions.

The Committee Members

Ramaiah : Age 46. You are quite close to the Minister. You believe in concentrating on your own gains and right now, doing as the minister says is the way best way out. You'd like to convince your other friends too that changing some of the players would be good for their career.

Shekhar : Age 48. You have been a player yourself in the past. You know that such phases can come in the career of any player. You have always stood by correctness and believed in professional ethics. You are a strong and determined person and you are determined not to let the play ground become the arena of politics.

The Coach (Chandra Shekar) : Age 42. You have been the coach of the team for the past one-year. You were brought in an attempt to boost up the dwindling performance of the team. You have dealt with similar situations before and you feel a little time and moral boost can solve the problem. You have always seen tremendous potential in each of the team members. Moreover, they are heavily dependent on one another. You feel that removing any of them now will only make the situation worse.

Role-Play – 8

The Situation : The owner of a company that produces automobiles and spare parts had ordered for software to regularize and streamline his business. He had asked one of the upcoming software companies to do the work. For some reason, however, the company could not supply the software in time. They extended the deadline much beyond the stipulated time. Finally when they were ready with the product the owner, Subhash Reddy found faults and refused to buy it. This is a situation where the representatives of the Software Company have come to meet Mr. Reddy and find a solution to the problem. The following is a rough sketch of the characters.

Role Descriptions

Mr Subhash Reddy : You are the owner of a reputed automobile company. You have come up the hard way in life and you believe in hard work and punctuality. You had ordered the software for the following tasks:

1. Keeping an updated record of your orders and supplies
2. Maintaining the employees profile.
3. Documenting the production procedure etc and regularizing it for later production.

You are irritated because the Software Company could not give you the product in time. In the meanwhile someone you know has got you software from abroad at a much cheaper price and now you do not require this product. The problem is that you had forgotten to cancel this order and earlier when they wanted an extension you had given it. You want to quote the delay as reason and refuse to take the software. Financially too you are under slight pressure but you cannot reveal this to the company.

The Representatives

Dr Anuradha : You are the head of the team that has produced the software. You know that there has been a delay but that was primarily because your team was burdened with too much work. You are efficient but you are also known to have a quick temper. You feel that Mr. Reddy is being unfair. He is known for his arrogance and unfair deals and you are determined to give him a piece of your mind.

Ajay : You are a shrewd and experienced marketing man. You are sure that the software can be marketed elsewhere if not here. But this was a good deal and you want to give it one good try. You also feel that it is important to maintain good relationships with Mr. Reddy for future deals. You don't mind even playing up to him if it means a long-term benefit.

Sudha : You are a young girl of 25 who has just joined the company. The company belongs to your father and of late because of his bad health you have had to take up more and more responsibilities. You have completed your M.B.A from a reputed university. And you are trying to find a solution to the problem. Your senior collogues tell you that you need to gain experience to be more effective. You are beginning to feel that all your study has been only theoretical. It hasn't helped you much in solving real problems.

NEGOTIATION SKILLS

> *Let us never negotiate out of fear.*
> *But let us never fear to negotiate.*
> *—J.F. Kennedy*

Speech as Negotiation

Negotiation is one of the chief activities in any speech situation. We commonly negotiate for discounts, prices, timings or options. In organizations again it is one of the most commonly needed skills that forms part of day-to-day communication. The outcome of a negotiation situation depends entirely on the values, attitudes, personal beliefs and emotions of the people involved in the situation.

Negotiation Styles and Their Contexts:

Depending on the context and the people involved in negotiation, we adopt different kinds of styles for optimum effect. The following are some of the negotiation styles we generally use:

> *One day, a small boy tried to lift a heavy stone, but could not move it. His father, watching, finally said, "Are you sure you are using all your strength?" "Yes I am!" the boy cried. "No you're not," said the father, "you haven't asked me to help you."*
>
> *—Internet*

	Negotiation styles	To be used when
1	**Collaborating**: a cooperative approach that stresses on win-win stance. Requires creative problem solving.	• Issues are very important and involve long term relationship. • Commitment from both sides is necessary to make things work.
2	**Compromising**: Here, both parties accept some amount of win and loss.The objective is to find some acceptable solution that both parties will accept	• There is only this alternative to no solution. • You need short term settlements. • Long term interests of both have to be kept in mind.
3	**Controlling**: Here, one ensures that his personal goals are met whatever be the consequences. It is a power-oriented approach where the only aim is to win.	• The action will be resented, but the issue is very important. • Quick action is vital and you believe that you are right
4	**Accommodating**: The first priority here is on conflict avoidance. It can be seen as Lose-lose or Lose-win approach that allows the other party to win	• You are in a weak position and you wish to minimize lose • Harmony and stability are more important than winning.
5	**Avoiding**: This style is characterized by evasiveness and withdrawal from the issue at hand.	• You are buying time. • There are more important issues Pressurizing you.

As we see, different kinds of styles are used in different situations. There cannot ever be one style that is true of all situations. A successful negotiator will always judge the situation, the people involved, and decide on an appropriate style.

> *Mother to young son:*
> *"You 'll love school, my dear.*
> *Think of it- a whole new world to mess-up!"*
>
> *—Readers Digest*

Points to Remember

In a communication situation that needs negotiation, always remember:

- Empathetic listening is the key to good negotiation
- Accept corrections if necessary.
- Remember the long-term consequences of your actions.
- Deal with deadlocks carefully. Try alternative methods, seek help from the opponent or suggest the appointment of an arbitrator.
- Apologize if necessary. A statement like, 'I'm very sorry, I didn't realize, helps.
- Use humor and adjournment to dissipate tension.
- Emphasize on building an understanding and reaching an agreement together.

ACTIVITIES

8. Given below is a list of suggestions about how one is likely to react during negotiation. Read them carefully and try to categorize them under the different kinds of negotiating styles. Where do you think you stand?

   a. I change the subject to a neutral topic.
   b. I complain until I get my way.
   c. I threaten the other party.
   d. I give up some points in exchange for others.
   e. I sacrifice my interests for the relationship.
   f. I try to convince the other person of the logic of my arguments.
   g. I postpone discussing the issue.
   h. I fight it out physically.
   i. I listen to the others' feelings.
   j. I look for a middle ground.
   k. I avoid hurting others feelings.
   l. I give up some points in exchange for others.
   m. I try to find out specifically what we agree and disagree on.

[Sometimes you may feel that a particular stance overlaps two or more negotiation styles. In such a situation try to create contexts and see when the stance can be said to be part of a particular style.]

9. Given below are a list of situations. See what style of negotiation will fit in best.

   a. You are seeking a promotion and you feel that your contribution to the organization warrants this. You have to negotiate this with your boss.
   b. You are the H.R chief in a company and you feel that the trade union leader is unnecessarily spreading ill will about the management. You have been given the responsibility of negotiating with him.
   c. You are the head of a software engineering team. You have been instructed by your boss to convince your team members and finish the work in half the time previously allotted.

d. Your brother wants to get married to a good friend of yours who you know is a drug addict. Neither you nor your family is happy with this decision. But you also feel that they can be given a chance. A family meeting has been called for and you have to negotiate the situation.

e. The municipal authorities will demolish your house, along with the entire colony for the construction of a flyover bridge .You have been chosen by your colony members to negotiate the settlement amount with the government.

10. List out ten different situations where you have to negotiate in your everyday life. Note them down and see what factors contribute to your success and failure in each.

11. **Case-Study**

Pick up two students from the group or divide the whole group into two and ask the groups to find their spokespersons. Read out the first part of the case study to the entire class. Then, distribute the copies of the two character descriptions given below (the character description of Dr. Subramanyam and Prof. Peter George) among the members of the two groups. Take care to see that they do not read the other group's sheet. Let the discussion begin after both the groups have read the sheets and discussed the problem with their individual team members.

Dr Subramanyam is the head of a research team that is experimenting on a new kind of preventive medicine. Prof. Peter George works at a University where the entire department of Chemistry is involved in developing a solution that can be sprayed in the atmosphere to minimize the negative effects of ozone depletion. Both the scientists belong to neighboring countries.

In Dr Subramanyam's country, unfortunately, there has been an epidemic outbreak of a new kind of disease that is affecting only children and if not contained it can kill thousands of them. Dr Subramanyam and his team have been asked to work on a war footing to produce a cure.

Prof. George too is in a high-pressure situation. Ozone depletion in his country has reached an all time high and it has been found that unless something is not done urgently it will affect a large part of his country and can cause deadly diseases in thousands of organisms. Prof. George has been directed to find a solution to the problem.

Dr Subramanyam: You (Dr Subramanyam) are the head of a research team that has been working for the past two years on a preventive medicine. After trying synthetic medicines your team has gone on to try natural products and now they have zeroed in on the extract of the bark of a rare kind of herbal plant found only in the disputed border area adjacent to your neighboring country. The plant takes four years to mature, only after which the bark can be used. The outbreak of the present epidemic has been devastating on the psyche of the country. With thousands of children on the brink of death and thousands more likely to be affected, the whole country is in the grips of panic. For you, too, it is a personal crisis because your only grandson has been affected and your daughter is distraught with sorrow.

If the team gets the barks immediately, they will need a week to experiment and finalize the formula. There is no time to grow the plant now and there is just about enough time to save the children presently affected and prevent more children from being affected.

The only problem in beginning the project is in procuring the plants. The disputed area has been taken over by an international peace keeping body and you cannot get access to the plant without prior permission. You have heard that the delay in giving permission is because of pressure from the neighboring country. They too have an immediate crisis and the same plant is necessary to help them out of the situation. The international body feels that both the countries need the same plant and the amount of raw material necessary will exhaust all the available plants. As a peacekeeping body they do not want to take responsibility of deciding in anyone's favor. They have asked you and your team to come on to the negotiation table and settle the problem agreeably. Since there is no time to grow the plant now you are convinced that you have to convince the other party, get the plants, and save the children.

Prof. Peter George: You (Prof. Peter George) are the head of a research team in a reputed university. Your department has received a special international grant to find a solution to ozone depletion. And you have been working on the problem for the last five years and your research expenditure has overshot the estimation. But you have been gracefully granted the extra amount too from the national fund .Now your country faces a crisis. The ozone depletion has suddenly worsened and is beginning to affect all the people who already have delicate health. Your research, after a lot of search, has shown that the liquid extract of the leaf of a plant can be used for the purpose. After mixing it with some other solution it can be sprayed in the upper atmosphere and the ozone depletion will be minimized for the next ten years. The rare plant, however, is not available in your country. As far as your knowledge goes, it is available in only one place-- the disputed area between your and your neighboring country. The area is in the control of an international peace keeping body that prevents all kinds of operation in the area. You have come to know from reliable sources that the same plant is urgently necessary in your neighboring country too. And since their need too is urgent they refuse to grant you the permission to use the plant.

The concessions you have already availed have made your position very delicate. You have assured the government of your country that you are on the verge of finding a solution to the problem. And now, unless you are able to deliver the formula, your career and that of your colleagues will be in jeopardy. Moreover the sudden threat to the lives of all the people who do not have good resistance has posed a new problem. The ozone depletion has to be immediately checked or within months yours will become a nation of sick people. The radiation has also proved to have adverse effect on the mental development of children. After procuring the plants you will need to work on it for about a couple of months to give the sprayable solution. Since there is no time to grow the plant now, you are convinced that for yourself, your colleagues and for your countrymen you have to clinch the deal and procure the plants immediately and finalize everything at the earliest. After you putting a lot of pressure the international body has agreed to allow you to negotiate with the representatives of the neighboring country.

Conduct the discussion at the negotiation style.

After the discussion, discuss the way the negotiation process was conducted. What factors contributed to their arriving at or not arriving at an agreement? Also discuss the body language and attitude of the discussants.

SUMMARY

- Effective speaking forms the crux of communicative skills: it is striking the right attitude and the right language at the right place.
- Researching the subject, selecting the content, planning for the talk is vital before a presentation.
- Remember to cut up your presentation sessions into short and brilliant sections. This will ensure audience attention.
- An effective presenter should prefer 'social space', avoid 'public space' and use body, eye, intonation effectively.
- Good visual aids can immensely improve the quality of presentation.
- Effective group behaviour is largely responsible for a good group discussion.
- Equitable distribution of participation leads to effective group discussion.
- It is important to network one's ideas, build up the argument and give the discussion a definite direction.
- Good listening forms an important part of group discussion.
- Negotiation is one of our commonest activities during day to day communication.
- Collaborating, compromising, controlling, accommodating, avoiding are different styles of negotiation.
- The outcome of an act of negotiation is dependent on the values, attitudes, and personal beliefs of the people negotiating.

REVIEW QUESTIONS

1. What are the different factors one has to be sensitive about to become an effective speaker?
2. What are the different speech styles?
3. During a presentation, how does a speaker ensure audience involvement?
4. What is "attention curve"?
5. What preparations are necessary before giving a presentation?
6. What are the factors crucial to an effective group discussion?
7. What is the importance of turn taking during a group discussion?
8. Define the concept of negotiation.
9. Mention some important factors that determine the success of negotiation.

CHAPTER 5

Reading Skills

> *Learning is a means of thinking with another person's mind; it forces you to stretch your own.*
>
> *— Charles Scribner*

Chapter Objectives

This chapter aims to present the basic principles of reading. Its objective is to familiarize the reader with the mechanics of the reading process and its connection with the process of cognition. It also aims to introduce different types of reading, and the situations they can be used in. The chapter also lists out certain techniques like chunking and silent reading that enable fast reading and draws attention to the SQ3R technique.

Very much like the skills of creativity, mind-mapping or negotiation, reading is a skill intimately connected with thinking. It is not simply about gathering meaning by understanding the structures and the vocabulary but also about cognitively processing the ideas, connecting them with one another and finally constructing a meaning from what is 'given'.

EYE MOVEMENTS AND CHUNKING

When we read anything, we actually rest the eye at places and take in a certain amount of material at a time. We therefore read in jerks. Our eyes can only take in information when they are stopped. So we proceed in a move-stop-read-move-stop-read movement. This can be observed if we look at the eye movement of a person who is reading. The moving and stopping activity of the eye takes about two seconds every time. So the lesser number of movements we have, the lesser time we take to read a certain material. The secret of quick and effective reading is to minimize the number of stops and maximize the number of words one sees at each stop.

> *Reading is to the mind what exercise is to the body.*
>
> *—Joseph Addison*

Activity

1. Here is a paragraph given below. Check the time you take to read it and also pay attention to the number of eye movements you make in every line.

   P.C. Mahalanobis is the genius who was primarily responsible for putting India in the Statistical world map. He was a professor of physics at the Presidency College, Calcutta in the early part of the twentieth century–a time when Statistics was nowhere regarded as an independent subject. He led an intellectual movement that firmly established Statistics as an independent discipline. He was also instrumental in the establishment of the Indian Statistical Institute, which spearheaded the growth and development of the discipline in the subcontinent.

Many of you would have taken in chunks as marked below.

P.C. Mahalanobis	is the genius	who was primarily	responsible for	putting India
in the	Statistical world map.	He was a professor	of physics	at the	Presidency
college,	Calcutta in the	early part of the twentieth century–	a time	when
Statistics was	nowhere regarded	as an	independent subject.	He led an
intellectual movement	that firmly	established	Statistics	as an independent
discipline.	He was also	instrumental	in the establishment of the	Indian
Statistical Institute,	which spearheaded	the growth	and development	of the
discipline	in the	subcontinent.

Consciously, try to reduce it to around three stops every line as shown below and try to check the time again. Try to position your eye roughly at the centre of every chunk and extend your vision to the right and the left.

| P.C. Mahalanobis is the genius | who was | primarily responsible for | putting
India	in the Statistical world map.	He was a professor of physics	at the
Presidency College, Calcutta	in the early part of the twentieth century–	a time,
when Statistics was	nowhere regarded as an	independent subject.	He led
an intellectual movement	that firmly established Statistics	as an independent
discipline	He was also	instrumental in the establishment	of the Indian Statistical

Institute,	which spearheaded the growth and development	of the discipline in the

subcontinent.

This would certainly reduce the time you take to read. If it does not happen the first time, you can definitely do it with some practice.

Points to Remember

To increase the speed of reading, you should:

- stop looking at every word, one at a time.
- try to take in words in groups,
- deliberately try to increase the number of words taken in at a time, thereby decreasing the number of times w e stop to make meaning of what we are reading.

SPEED READING

With ever-increasing literature in every field, having a good reading speed is inevitable if one is to be fairly acquainted with the developments in the various disciplines of study. A good reading speed is something one cultivates with constant practice. Given below are some of the ways in which this can be done.

1. Reading selectively

Learning to read well is also about learning what not to read. To read fast and to read efficiently, you have to learn what to skip, ignore and choose what to focus on. It is important, therefore, to immediately recognize the key words, identify repetitions or restatements, avoid the structure words or the fillers, make inferences about the meaning and focus on the relevant chunks. This process implicitly involves making a quick mental map of the material to be read so that one can organize the information mentally, get back to the relevant portions when necessary and return easily to the place one had moved from. Effective reading, thus, is not necessarily the act of reading word by word, line after line. More often than not, it is a to--and –fro activity where you move foreword, come back, and proceed once again.

Points to Remember

Important factors while reading are:

- Focusing on key words and ignoring filler words.
- Skipping what you already know.
- Skipping material that does not apply to you.
- Skipping material that seems particularly confusing and coming back to it if necessary after reading other sections

2. Reading without subvocalising.

Like the point discussed above, this is another way of increasing the speed of reading. Often while reading silently, we tend to pronounce the words in our mind. This slows down our reading speed, almost as if we were reading things aloud. This also forces us to read every word, making it impossible to skip or infer. It is necessary, therefore, to consciously stop this method of reading and practice reading and processing with the eyes only. To do this effectively, one has to:

- Move from the "see → say (sub-vocally) → understand" to just "see → understand".
- Practise taking in semantic chunks at a time and practice what is called "mental reading".

This habit of taking in larger chunks of material at a time also helps in pulling out words from different lines and synthesizing meaning. Reading, hence, can become simultaneous with thinking, a way of structuring meaning. One of the ways of adapting to multiple reading speeds and styles is to practise to read with the eyes only.

Given below is a picture of three eye-movement patterns showing how reading can be chunked in larger and larger units in the same line and sometimes across lines.

Source: http://braindance.com/bdiread2.htm

Activities

2. Very quickly gloss through the small passage given below in about 20 seconds. Next, underline the words and sentences you focused on, to quickly get a sense of the meaning. Compare it with the shaded parts of the passage given below.

The study of Mathematics is essential for all children as it has widely ranging application in day-to-day life. The learning of Mathematics at the primary stage provides the child with the basic mathematical concepts and skills needed to tackle real-life problems. It helps to cultivate thinking and reasoning abilities, thereby, strengthens the intellectual underpinnings of social interaction.

The study of Mathematics is essential for all children as it has wide-ranging application in day-to-day life. The learning of Mathematics at the primary stage provides the child with the basic mathematical concepts and skills needed to tackle real-life problems. It helps to cultivate thinking and reasoning abilities, thereby strengthening the intellectual underpinnings of social interaction.

It is quite possible that you too would have swept your eyes across the key words marked to get a sense of the meaning. One has to do this while reading selectively and effectively at the same time.

3. Again, read through the paragraph given below and make a note of the semantic chunks that one has to focus on to get a general feel of the content. After reading, identify 10 most important words or phrases that would help one to construct the meaning.

The history of the ancient Egyptian state is one of successive periods of unification and fragmentation. Counterbalancing this, is a pattern of civilization – characterized by such features as the use of writing, an organized system of religion and divine kinship, and dependence on the annual Nile floods for the fertility of the land—which links the different periods together through a span of 3000 years.

With the increasing amount of available information in every field, it has become necessary, in fact inevitable, that selects are information for reading. One way of doing this is to skim through most of the material, select what is important and focus on the necessary information. But this may not be possible in all reading situations. At times one has to read very closely, focus on every word, even read in-between the lines (inferential reading) to get the meaning implied.

4. Read the paragraph given below and answer the questions given.

India is a land of ancient civilization that had made fundamental contributions to human knowledge in the ancient past. Even the discovery of the positional number system with the accompanying concept of 'zero' is attributed to ancient India. We already have a population of more than a billion and may soon become the most populous nation on Earth. Should we rest content with our ancient heritage and keep repeating that our contribution to knowledge is 'zero'? Should we continue to be mere borrowers and users of modern knowledge and modern scientific technology? When do we give back? When do we become creators of fundamental knowledge?

1. According to the paragraph, the concept 'zero' is:
   a. a part of the positional number system.
   b. a product of the positional number system.
   c. connected to the positional number system.
   d. None of the above.

2. Which of the statemets below is most correct?
   a. The number system is the only contribution India has made to the world.
   b. Ancient India has made many contributions to the world.
   c. The contribution of 'zero' is the most valuable gift of India to the world.
   d. None of the above.
3. "Should we … keep repeating that our contribution to knowledge is 'zero'?" what does the word 'zero' mean here? Does it have more than one meaning?
4. "We already have a population of more than a billion and may soon become the most populous nation on Earth". Why has this information been given here? How is it linked with the questions that have been posed towards the end of the paragraph?
5. What is the meaning of fundamental knowledge? Are there some forms of knowledge that are not 'fundamental'? Given below are some knowledge fields. How many among them can you consider to be fundamental?
   a. Astronomical discoveries.
   b. Different applications of principles of physics in constructions field.
   c. Knowledge of herbal medicines.

> *The flood of print has turned reading into a process of gulping rather than savoring.*
> *—Raymond Chandler*

In the activities above, you would have inferred that there are various kinds of readings. Depending on the need and the type of material, we decide on a reading mode and proceed accordingly. Most often, we categorize the speed of reading into the following two types:

1. Skimming
2. Scanning

Skimming is the skill of quickly going through a lot of material to gather the main points. It is necessary to skim when we are trying to get a general idea or sorting out information, trying to check what might be important or even rejecting information as irrelevant. It includes a quick noting of conspicuous text like headings, subheadings, bold or italicized words.

Scanning is done to locate a particular piece of information while going through a general document. It is also done to interpret or reconstruct a pattern in the data presented. We scan while closely examining schedules, meeting plans or train/flight timings. We also scan the index or the bibliographical references when we need certain information. This kind of activity needs close reading and detailed look at information. Sometimes, it might even involve comparison of facts and data presented.

The categorization done above is based on reading speed. There can also be a categorization on the basis of the kind of reading. They are:

1. Extensive reading
2. Intensive reading

Extensive reading does not focus on particular details but look at ideas and fact patterns. One does extensive reading while looking for a research problem, getting a grasp of work done in a particular field, looking at the trends of publication at a particular time or even reading for pleasure.

Very different from this, is **intensive reading**. Like scanning, this involves reading to get specific information. But while scanning generally refers to the close checking of lists and tables, this pertains to reading texts and longer material. Intensive reading is done to closely follow a line of argument, read in-between the lines and look for implied positions. Preparation for teaching very often involves intensive reading. It requires paying attention to minute details and understanding detailed facts. Reading for research activity, too, requires intensive reading.

ACTIVITIES

5. Look at the following data and state the general theme of the writing. What kind of reading would you like to adopt for this?

   In the summer of 1789 the most powerful monarchy in Europe was overthrown by a popular revolution in favour of constitutional rule. There followed ten years of political turmoil as moderate parliamentarians and radical republicans fought over the political future of France. The many different ideals presented by the revolution were impossible to reconcile, and in 1799 a new era of authoritarian rule began under the dashing revolutionary general, Napoleon Bonaparte.

6. Given below, are some situations in which we do reading. State what kind of reading you would do in these situations? Remember, depending on your need you would read the same material differently at different points of time. So answer this, given your present condition and necessities.

   - The TV guide for the week
   - An English grammar book
   - An article in a popular magazine about the famous actors of yesteryears.
   - The weather report in your local newspaper
   - A bus timetable
   - A fax that has come for you at the college office.
   - An email or letter from your best friend
   - A recipe
   - A short story by your favourite author

THE SQ3R METHOD

The SQ3R method was put together by Francis Robinson in 1970 as an active and effective method of reading. It is one of the methods best known for reading faster and retaining more. SQ3R stands for the steps in reading: survey, question, read, recite, review. The following are the steps in detail.

Survey

Before reading, survey the material. Look through the main and the sub-headings and try to get an overview. Skim the sections and read the final summary paragraph to get an idea of what the chapter is about. Pay attention to introductory and summary paragraphs and references.

Question

Ask yourself what this chapter is about: What is the question the reading material is trying to answer? Repeat this process with each subsequent section, turning each heading into a question. Asking questions focuses your concentration on what you need to learn or get out of your reading. This step requires conscious effort that it leads to active reading, the best way to retain written material.

Read

Read one section at a time, looking for the answer to the question proposed by the heading. This is active reading and requires concentration. If you finish the section and have not answered the question, reread it. Think while reading. Consider what the author is trying to say, and think about how you can use that information.

Recite

Once you have read an initial section, write down (sometimes in the margins of the book itself) a key phrase that sums up the major point of the section and answers the question. Here, it is important to use your own words, and make your own connections. Research shows that we remember our own (active) connections better than the ones already provided. At this stage, writing down the answers to your questions helps in retaining and consolidating your understanding.

> *When you read a classic, you do not see more in a book than you did before; you see more in you than there was before.*
>
> —*Cliff Fadiman*

Review

After reading the entire material, again ask yourself the questions that you have identified right at the beginning. Review your notes for an overview of the chapter. Consider how it fits with what you already know. Think of the significance it has in your general learning scheme.

SUMMARY

- Reading is intimately connected with thinking.
- Taking in more material with every eye movement increases the speed of
- reading.
- Reading selectively and reading without subvocalisation are important.
- Learning skimming, scanning, intensive and extensive reading help us adjust the speed and manner of reading according to the material to be read.
- The SQ3R technique is very effective when we need to do active reading for retention as well as understanding.

REVIEW QUESTIONS

1. Give a brief account of the different kinds of reading.
2. Look around you and make a list of three kinds of material you use for the different kinds of readings.
3. What are the different methods one can adopt to practice fast reading?
4. Describe the SQ3R technique as an active reading process.

CHAPTER 6

Writing Skills

> *'The thing that gives me and has always given me the most happiness in life, is writing. The mind celebrates a little triumph every time it formulates a thought'*
>
> *—Emerson*

Chapter Objectives

This chapter aims to familiarize students with some important forms of writing like paragraph writing, précis writing, instructional writing, summary and abstract writing. It also discusses the factors critical for good writing like the processes of writing and sequencing. Further, it brings into focus the important skill of note-making.

> *A drop of ink makes thousands, perhaps millions think.*
>
> *—Byron*

THE BASICS OF WRITING

> "All language demonstrates three kinds of excellence: correctness, precision, and elegance. Language also has the same number of faults, and these are the opposites of the qualities just mentioned"
>
> —Quintilian

We can add a few more factors to those mentioned by Quintilian. All these factors contribute to clear, fluent, and effective writing:

- Purpose: The reason for writing;
- Audience: The readers;
- Writer's Process: Getting ideas, getting started, writing drafts, revising;
- Mechanics: Handwriting, spelling, punctuation;
- Grammar: Rules of verb, agreement, articles;
- Syntax: Sentence structure, stylistic choice;
- Content: Relevance, clarity, originality, logic;
- Word choice: Vocabulary, idioms, tone;
- Organization: Paragraphs, topic and support sentences, cohesion and unity;

Writing can also be seen as the skill of communicating with either the 'now' audience/readers or 'later' audience/readers through the medium of paper/ screen/display. The effectiveness of writing depends among other factors, primarily on clarity of expression. In addition to clarity of expression, an effective writer also needs to have a good style that is appropriate and unambiguous.

The act of writing anything involves an effort to express ideas through the constant use of eyes. The hand and the brain and comprises a unique way of reinforcing learning and discovering new ways of expressing ideas. Reading is a skill that is very closely related to writing. Every act of writing is seen as an act of reading, and reading itself is a kind of writing. From one perspective, writing is the way in which we evaluate or express reading skill, and reading is the way in which we evaluate writing skill. The close relationship between writing and thinking on one hand and writing and reading on the other makes writing a valuable part of any language learning or skill-acquiring programme.

All effective writers avoid ambiguity. This can be ensured by being careful about the word order. Ambiguous sentences can be entertaining in their double meaning, but can mar precision in writing.

Activity

1. Look at the following sentences and see what meanings can emerge from each and change the word order to remove the ambiguity in them:

   a. I saw our principal walking through the window.
   b. Flying planes can be dangerous.
   c. Visiting relatives can be boring.
   d. He went to the market to sell the house along with his wife.
   e. Available–4-bedroom house with lovely trees being constructed by a foreign architect.

THE PROCESS OF WRITING

Any kind of writing can be seen as a process in three stages:

- Pre-writing
- Writing
- Post-writing.

But it should also be borne in mind that writing is too complicated a process to be broken up into three neat stages. It is and has to be full of overlaps. It is recursive–often starts, stops, loops backward and goes forward again. These stages can be seen as rough break points that are to be kept as guiding principles while writing.

Pre-Writing

This is probably the most crucial stage in the writing process. It involves forming a thesis statement and an outline. At this point, one has to formulate a clear idea about the purpose of writing, the audience and generate ideas about the kind of information one wants to pass on.

Some of the commonly used techniques during this stage are brainstorming, clustering and clubbing of ideas. Techniques like mind-mapping or using any other way of branching and organizing would help in sequencing and forming idea clusters in the mind. If you have been able to arrive at rough headings and subheadings, it is important to check out the following points:

- Do each of your headings describe one of your central ideas?
- Have you been able to add on further sub-points to each? Can you have one paragraph on each of the main headings?
- Are all the headings free of overlapping ideas?
- Are the headings logically related to one another?

ACTIVITY

2. Take up any one of the given topics, try to brain storm around them and put the ideas in any order you feel convenient.

    a. Creating a sustainable environment.
    b. The role of the media in creating stereotypes.

Writing

The next stage is the actual process of writing, elaborating and filling out the frame prepared in the prewriting stage. The important concerns here are, dividing the writing into the **introduction**, the **body**, and the **conclusion**.

In the **introduction**, it is important to:

- introduce the subject
- set the direction of the writing
- capture the imagination of the reader

In the **body** of the writing, one has to pay attention to the sequencing of ideas, the logicality and coherence of presentation and a strong sense of direction. The body of the writing should contain at least one fully developed paragraph about each of the central ideas listed out in the prewriting phase. They could follow any logic in the order of presentation (either from the least important to the most important or any other).

In presenting the main and the sub-points, you could either proceed from the general to the specific or from the specific to the general. Very often people prefer moving from the general to the specific. It is called the funnel method of presentation.

The **conclusion** is largely responsible for giving the reader a sense of completion, a feel of tying up the loose ends. It could be a summary or an evaluation of the ideas previously presented. A conclusion is largely responsible for reinforcing and concretizing the argument of the writing. It also makes clear the writer's position on the issue being discussed. One has to be careful therefore about the way it is worded.

Post-Writing

This is the third and last step in the writing process. It includes the task of rereading the paper to see what revisions might need to be made. This often means more than just proofreading for minor mechanical errors, such as spelling and punctuation. A good writer will always be critical of his/her writing at this stage. Here, it is important to:

- be objective
- keep the purpose of writing that you had developed in the prewriting stage.
- keep in mind the audience and their expectations.

Along with these factors, you have to focus on the appropriate formatting paying attention to the space, margin and font. Finally, before submitting, it is important to once again check for spelling, punctuation, omissions or any other careless mistakes.

Pre-writing	Writing	Post-writing
Time taken (25%)	Time taken (25%)	Time taken (50%)
Planning generating information	Structuring, organizing that information in an acceptable fashion	Revising, polishing the draft.
Techniques used: Brain storming, Branching, Flow-charting	Techniques used: Comparison – Contrast Problem – Solving	Techniques used: Adding, Deleting, Simplifying, Erasing errors

By three methods we may learn technical writing: First by education, which is noblest; second by methodology, which is easiest; and third by planting your butt in a chair and pecking out the damn document, which is the bitterest.

—*Andrew Plato*

Activity

3. a. Given below are some points that could probably be generated while discussing the first topic in Activity 2. Put them in the form of main and sub-points in the way indicated. Add more points if you can. Develop it into a well-organized and complete piece of writing.

Creating a sustainable environment.

- an environment that can maintain itself.
- checking industrial growth.
- controlling harmful gasses/ chemicals/ industry waste.
- monitoring the ozone depletion, glacier melt.
- saving greenery/preserving ground water level.
- checking the contamination of sea water / protecting sea life.
- water harvesting /plastic use/ land fills
- saving forests/ planting new forests,

b. Follow the same method–generate ideas, put them in the form of points and sub-points–and write an essay on the second topic.

PARAGRAPH

A paragraph can be seen as the basic unit of connected writing. A good paragraph is one that carries a single idea and where the argument or the idea itself is organized logically and presented coherently. The **topic sentence**, which is the main sentence of a paragraph, conveys the central theme or idea. It is generally written in the beginning of a paragraph. However, it can be written either in the middle or at the end of the paragraph. If mentioned at the beginning, the sentences after the topic sentence are explications or extensions of it. But if it comes at the end, the sentences generally lead the reader to the topic sentence. Sometimes a topic sentence can be implicit within a paragraph when it is not written. In such a situation, the other sentences of a paragraph should lend support and substantiate the topic sentence.

One of the skills needed to develop or organize a paragraph well is appropriate sequencing. While writing a paragraph, one has to keep in mind the logical ordering of ideas.

Sequencing

Sequencing is the skill of organizing a textual material, deciding the priority, the focus of the different points and consequently the order in which they should be presented in a paragraph. Sequencing is also about linking up ideas and concepts and using the correct linkers to show the relationship they have with one another. Linkers are essentially thought connectives. Some of the widely used linkers are as follows:

In addition to; further; moreover; apart from; although; however; though; in spite of; whereas; on the contrary; for example; for instance; thus; such; in addition; furthermore; then; in this case; indeed; surely; above all; certainly; in the same way; on the other hand; in contrast; whereas; instead; similarly;

more importantly; additionally; in the same way, because, especially, then, of course, fortunately, before, after, besides, well, in other words, even; but etc.

In addition to using linkers, sequencing also involves using the appropriate words and making the writing brief, crisp and clear. Attempt the following activities.

ACTIVITIES

4. The following passages are in jumbled order. Read them carefully and sequence the sentences to make the passages logical and coherent. The first sentence (S1) of the passages has been given at the beginning:

S1. In every society, there has always been some or the other form of education, but not necessarily schools.

a. In some places, pupils trained under a master craftsman in whose household he lived with other trainees: here they were taught the skills, the ethics, the principles, rules and customs of business.
b. In certain ways, they were better educated than their descendants today.
c. Two hundred years ago, very few people in India went to schools or colleges.
d. The peasant in the village learnt the arts and skills of farming, together with a lot of traditional wisdom about the earth, the sky and living things.
e. But this lack of institutional education did not mean that people were not as or in cases better educated than we are today.

S1. A detergent is one of a class of chemicals used for cleansing purposes.

a. They lather easily even in hard water.
b. Soapless detergents differ from soapy detergents in that they contain no animal fats.
c. The earliest known detergent was soap.
d. However, detergent foam in sewage works and rivers can cause problems. Therefore detergents now are of a soft or biodegradable kind, which can broken down by bacteria.
e. But there is also a soapless detergent as distinct from soap.

5. The sentences in the following passage are all jumbled. Rearrange them in a logical sequence. You may divide the passage into as many paragraphs as you consider correct.

It is not really a question, in spite of the question mark. The complications begin, of course, when you must use two rules at once. A casual, yet correct introduction often involves merely saying the two names, or formally saying – this is..., I'd like you to meet..., may I introduce...? Children rise when introduced: a lady rises to meet a well-known person or an elderly woman: and a gentleman rises for all occasions. Introductions form an essential part of social intercourse and have their own etiquette. When one says – how do you do? – The reply is the same – how do you do? – though some people prefer – glad to meet you – either instead, or additionally. They are governed by three simple rules. Introductions also have their own set phrases. In such cases, use a mixture of respect and common sense. The normal reply is – how d'you do? – Introduce the younger person to the older; introduce the less important person to the more important; and introduce the male to the female. For example, introduce a very young girl to a much older man.

6. Rewrite the following text appropriately by cutting out the redundant words and rephrasing the sentences.

   "In many parts of the world there are large areas of potentially good land that cannot be used for agriculture because they are either too dry or too wet. But by engineering skill–devising efficient methods of irrigation, drainage, or flood control–and improved agriculture, it is often possible to reclaim such land."

7. You are taking a written test for promotion in the company. Your first question is a badly written passage. You are expected to correct and edit the passage appropriately:

   "Research in solar energy is as sophisticated and intensive in India as anywhere else in the world. For ten months with the year, six to eight hours in a day, much of India receives high intensity, fairly uniform and equal sunshine. The India government has therein given priority at six projects in the development of solar energy."

8. Read the following passage and fill in the blanks appropriately. Choose from the list of words given:

   I like everything about my motorbike, ______ its colour and speed. ______ recently, its pickup has decreased. I don't know why. I'll take it to a mechanic ______ it becomes worse. Most mechanics these days are undependable, but ______ my mechanic is reliable ______ being economical. Actually, ______ I bought my bike in 1998, I have been having a good driving time, except ______ for an occasional problem here and there. It has been giving me a decent mileage. I feel the mileage will increase further in a few days' time, ______ I'm taking that extra little care of it. ______, I'll have to wait and see. But in the meantime, a friend of mine advised me to sell the bike and buy a different model. I don't think I'll do that, at least not in the near future. I will use it for five more years and ______ I'll sell it may be. But the idea is very disagreeable to me.

   List of words: *but , because, especially, then, of course ,fortunately, before, after, besides, well, in other words, even.*

9. The editor of a daily has invited readers to send their views on the following topics. You decide to write a paragraph on each. You may make use of the hints given:

   a. Privatization of higher education – less strain on government; better quality; sincerity and commitment; setting of value; placements
   b. Global warming – decreasing green cover; rapid urbanization and industrialization; afforestation; environment-friendly attitude and policies.

10. Read the following:

    Only exceptionally, people find a long journey by motorcar agreeable. Mostly a long non-stop journey by train is much more comfortable and enjoyable. The tracks, the coaches and the smooth though swift motion all add to the pleasantness of the train journey. Looking at the beautiful countryside from a window seat has its pleasures too. One can experience a sense of adventure together with a restful sleep. Modern railway travel is much more comfortable than in the past. It revives the childhood excitement of visual and imaginative delight.

    Now, write out a paragraph listing some advantages of motorcar journey over the train journey.

11. Here are two topic sentences given. Write out two separate paragraphs building on and supporting the topic sentences:

    a. The computer has made the business world more effective.
       Hints: saving of time and space; quickness in communication and trade.

    b. Motivation is important for better performance in business and industry.
       Hints: enthusiasm; alertness; self and company image; drive for prosperity.

Read the following:

The slogan of an anti-liquor campaign read: Alcohol kills you slowly.
The next day there was a rejoinder written below it: We are not in a hurry.

INSTRUCTIONAL WRITING

Instructional writing is the ability to write out instructions precisely and unambiguously to prospective readers. It generally has an overt or a covert style of putting across certain do's and don'ts in a given context. The most common form of instructional writing is found in different kinds of user manuals. The context can be commercial, academic, personal, social, investigative or any other in nature. In a world of increasing gadgets and equipment, instructional writing has become an essential skill to be learnt. One has to have written instructions to be able to use a camera, to prepare income tax returns, to program VCRs and to run machines. To write a piece of instruction, one has to do the following:

1. Define the goal clearly.
2. Determine the audience.
3. Decide the most acceptable sequence.
4. Use visual aids where necessary.

It is important for instructional writing to first define the end result. One has to be clear whether one wants the writing to help people understand the process involved in a gadget or simply to operate the gadget.

Determining the audience is equally important in instructional writing. One has to estimate the awareness level of the audience in the field, determine the kind of exposure they have had and then determine the kind of instructions to be given.

To determine the right sequence of presentation in instructional writing, one has to often foresee difficulties. One has to be able to think what can go wrong or where one can be confused at each step, give the solutions and accordingly determine the steps. The following is an example of the kind of instructions that are given to a beginner and to an experienced user respectively.

Logging instructions

Beginner:

1. Flip on the power switch. It is on the back of the terminal to the left.
2. Press the return key until ENTER CLASS appears on the screen.

Note that the computer has a 20-second time limit on the 6 instructions that follow. So you must move right along or you will have to restart.

3. Type in "3". The VAX is a class 3 option.
4. Press the return key. The computer will respond with "GO".
5 The Computer Will then Display "User Name" On the Screen.
6. Type in "SHREE GANESH". This is the training session user name.
7. Press the return key. "Password" will be displayed.
8. Press "ENTER" To Begin your work.

Advanced user:

Turn the terminal on.
Type return.
Type 3, Return.
Type, Return, Return.
Insert user name, Return.
Password.

Adapted from *Technical Report Writing Today* by Steven E. Pauley and David G. Riordan.

Activities

12. Your friend is planning to come to your house for dinner. As he is coming for the first time he is not sure of the exact location of your house. Write out a set of instructions mentioning the important landmarks. You may make use of hints like turn right, come straight, crossroads, water tank, hotel and phone booth.

13. The makers of ALL-CLEAR washing machines in your town urgently need to print out copies of the User Manual next week. You are engaged in writing a set of 5 instructions for the manual. Write them out clearly and precisely. Use the following hints:

    Running water tap; outlet pipe; select speed; soak facility; disconnect

14. You have been engaged by LUSTRE to write four instructions for their new hair dye product named L'HAIRBL. Make use of the following hints:

    Bowl, spoon, warm water, glove, comb

PRÉCIS WRITING

The word précis is adopted from French and signifies a 'summary' or an 'epitome'. Writing a summary of a text, whether it is a speech, a biography or any other, is a difficult and useful skill, demanding *concentration*, *comprehension* and *condensation*. After understanding a passage, we have to compress it. In compressing a passage, *selection*, *rejection* and *generalization* are often useful. The general practice has been to prepare the summary of a paragraph in about a third of the original. However, there is no fixed rule about this. You may also give a title to the summary.

Points to Remember

While writing a summary,

- separate the relevant points from the irrelevant ones
- arrange the ideas systematically in a logical sequence
- condense material in as short a form as possible
- avoid using the same words, phrases and expressions used in the text
- rephrase the material in simple language
- avoid comments, abbreviations, symbols and examples.

Look at the following example:

An old man came to Hyderabad for the first time. His daughter was employed in Hyderabad but she could not come to the station to receive him and take him home. The old man came out of the station. He spoke to a taxi driver, negotiated the charges and gave him the address. He got into the taxi; the taxi started but it moved at such incredible speed and in such an erratic manner that it almost collided with a van; after sometime, it was about to knock down a traffic personnel and collide with a lorry. The old man got scared. 'Take care', he cried out, terrified. Patting on the driver's shoulder to attract his attention, he said 'you are frightening me by this manner of driving. I am afraid I will be in the hospital instead of my daughter's house. I want you to be more careful. This is the first time I am going by car. I am very nervous". The driver, who was sweating by now said, 'I understand your nervousness; I sympathize with you. This is the first time I am driving a car!' (About 190 words)

Précis:

Their first attempt

A taxi hired by an old man who arrived in Hyderabad was traveling erratically at great speed. It narrowly missed hitting other vehicles. The terrified passenger requested the driver to slow down. He even said that it was the first time he was traveling by car. The driver expressed his sympathy and added that it was the first time he was driving a car! (64 words)

As far as possible, it is preferable to use your own words in making a summary. However, in the case of any important expression, the original words may be retained.

ACTIVITIES

15. Rewrite the following sentences, condensing them wherever possible without altering the meaning.

    Example: This is an engine-propelled road vehicle.
    This is a car/bike.

    1. It was not without substantial feelings of pleasure that I did the work for them.
    2. The little feathered creatures in the sanctuary made such sweet sounds!
    3. Aunts, uncles, grandparents, and cousins are generally helpful and useful.

16. Write a précis of the following passages, condensing them wherever possible without altering the meaning.

    First and foremost there are order and safety. If today I have a quarrel with another man, I do not get beaten merely because I am physically weaker and he can knock me down. I go to law, and the law will decide as fairly as it can between the two of us. Thus in disputes between man and man, right has taken the place of might. Moreover, the law protects me from robbery and violence. Nobody may come and break into my house, steal my goods or run off with my children. Of course there are burglars, but they are very rare, and the law punishes them whenever it catches them. It is difficult for us to realize how much this safety means. Without safety those higher activities of mankind which make up civilization could not go on. The inventor could not invent, the scientist find out or the artist make beautiful things. Hence order and safety, although they are not themselves civilization, are things without which civilization would be impossible. They are as necessary to our civilization as the air we breathe is to us; and we have grown so used to them that we do not notice them any more than we notice the air.

    (From Our Own Civilization, C E M. Joad, 209 words)

17. In democratic countries men and women are equal before the law and have a voice and choice in deciding how and by whom they shall be governed. But the sharing of money, which means the sharing of food grains and clothing and shelter and knowledge and other such essentials is still very unfair. On one hand some people live in luxury, on the

other hand, many have not even enough to eat and drink and wear. Even in the richest of the world's cities thousands of people live in miserable surroundings. There are many families of four, five or six persons who live in a single room; in this room they sleep and dress and wash and eat their meals; in this same room they are born, and in this same room they grow up and die. And they live like this not by choice or for fun, but because they are too poor to afford another room. It is, I think, clear that until and unless everyone gets his or her proper share of necessary and delightful things, our culture and civilization will be far from perfect. (188 words)

18. Yet another great defect of our civilization is that it does not know what to do with its knowledge. Science, as we have seen, has given us powers fit for the gods, yet we use them like small children. For example, we do not know how to manage our machines. Machines, as I have already explained, were made to be man's servants; yet he has grown so dependent on them that they are in a fair way to become his masters. Already most men spend most of their lives looking after and waiting upon machines. And the machines are very stern masters. They must be fed with coal, and given petrol to drink, and oil to wash with, and must be kept at the right temperature. And if they do not get their meals when they expect them, they grow sulky and refuse to work, or burst with rage, and blow up, and spread ruin and destruction all round them. So we have to wait upon them very attentively and do all that we can to keep them in a good temper. Already we find it difficult either to work or play without the machines, and a time may come when they will rule us altogether, just as we rule the animals.

    (From *Our Own Civilization*, C E M. Joad, 212 words)

ABSTRACT WRITING

> Abraham Lincoln always labored to keep his speeches short. He spoke a little more than two minutes at Gettysburg, some 50 minutes less than Bill Clinton did at the 1992 Democratic convention.

An abstract is a brief summary of the contents of a research report, article or presentation. Generally, the title and author(s) name are added to give it context. Traditionally, the abstract covers an Introduction, Methods, Results and Discussion (IMRD format) – in the shortest amount of space possible.

1. **The Introduction** typically describes the problem and its importance. It should include:

    a. Appropriate problem description.
    b. The specific questions you are going to answer.
    c. The purpose, motivation or relevance of your problem.
    d. What you hope to learn or achieve from your research.

2. **The Methods** are the framework, procedures, and tools for investigating your defined problem. As part of this, you need to:

   a. Summarize all the important information related to strategy and methodology
   b. Describe the analytical techniques used.

3. **The Results (or outcomes)** of your work should be concisely and objectively listed in a logical sequence. Here you should mention if:

   a. any comparisons are being made to existing ideas.
   b. you have developed software or hardware.
   c. you did a benchmark study and if it was appropriate to do so.

4. **The discussion (or conclusion)** offers an evaluation and interpretation of your findings and makes some suggestions about solutions to your stated problem. Can you make generalizations or projections of new insights into your scientific field? Are there any future improvements to consider?

The art of writing a good scientific abstract is to:

- Address the four key elements of the IMRD format using two or three well-constructed sentences per element.
- Use simple statements
- Use precise language
- Use well-known abbreviations to keep it short and simple

In today's world of fast business transactions, the skill to write precisely yet briefly has become a necessity. For effective writing, you have to avoid unnecessary details, roundabout expressions and come to the point directly. Look at the following excerpt from a letter:

> "We are in receipt of your esteemed letter of 5 September and in reply beg to state that we have conveyed the information to our head office in Calcutta. On hearing from them we shall be in a position to convey to you our exact and considered decision."

We can put the same information in the following form.

> "Thank you for your letter of 5 September. We have written to our head office in Calcutta and shall let you know our decision on hearing from them."

Sentences can be abridged sometimes by substituting words or phrases or removing ornamental and superfluous expressions. Read the following illustrations.

1. a. *His conduct was such that it could not be excused.*
   b. *His conduct was inexcusable.*

2. a. *When I began to write my first report it occurred to me that one method by which I could make it more effective would be to include a series of diagrams.*
   b. *While writing my first report, I realized that including diagrams will make it effective.*

Activities

19. Abridge the following sentences:
    a. He refused to accept the explanation given by the mechanic.
    b. I cannot do this unless I have the instruments, which are necessary for the work.
    c. After a thorough investigation was made, it was decided to make an exact estimate of the value of the property that was damaged.
    d. The immediate objectives or the short-term goals of any organization are present along with the necessity of having long-term goals or goals of longer range. Even while the incessant fight with daily crises like supervising, hiring, selling and profit making goes on, any forward-looking organization is also concerned with the company's general growth, reputation, and more importantly, creating and establishing an identity.
    e. An appropriate understanding of a plan by the people who are concerned with a company plays an essential part in its fulfillment. They should all be able to understand the main lines or the direction in which development is expected to proceed in the future. They should also be able to see the progress in different directions and establish the links necessary to relate to one another and see how it can be strengthened and how firmly establishing one area can help in strengthening other areas too.

20. A new employee of your office has just given in a clumsily written passage. As the chief editor, you need to edit it by making it concise and appropriate:

    The very need for food, fodder and fuel has increased tremendously in the last few recent decades. Trees are being cut down ruthlessly and mercilessly for non-forest purposes like agriculture, dams, roads, defence and transport. This has resulted in and led to depletion of forest wealth, which in towns is leading to floods, siltation of dams, water shortage, logging and also increaseing erosion of fertile soil.

21. Read the following passage and write out its abstract in about eighty words:

    One of our most difficult problems is what we call discipline, and it is really very complex. Actually, society feels that it must control or discipline the citizen, shape his mind according to certain religious, social, moral and economic patterns. Now, is discipline necessary at all? Please listen carefully, don't immediately say "yes" or "no". Most of us feel, especially while we are young, that there should be no discipline, that we should be allowed to do whatever we like, and we think, that is freedom. But merely to say that we should be free and so on has very little meaning without understanding the whole problem of discipline.

    The keen athlete is disciplining himself all the time, is he not? His joy in playing games and the very necessity to keep fit makes him go to bed early, refrain from smoking, eat the right food and generally observe the rules of good health. His discipline is not an imposition or a conflict, but a natural outcome of his enjoyment of athletics.

    Now, does discipline increase or decrease human energy? Human beings throughout the world, in every religion, in every school of philosophy, impose discipline on the mind, which implies control, resistance, adjustment and suppression—is all this necessary? If discipline brings about a greater output of human energy, is it very harmful, destructive? All of us have energy and the question is whether that energy through discipline can be made vital, rich and abundant, or whether discipline destroys whatever energy we have. I think this is the central issue.

22. In the following paragraph, can you eliminate more than 15 words?

"Edit ruthlessly. Somebody has said that words are a lot like inflated money—the more of them that you use, the less each one of them is worth. Right on. Go through your entire letter just as many times as it takes. Search out and annihilate all unnecessary words and sentences—even entire paragraphs."

Read the following:

A legal secretary once sent a letter to a client saying, "kindly attend at my office in order that we might execute the necessary documentation". After receiving the letter she called the office and reread the instructions to her with a question in her voice. "Does this mean I have to come in and sign some papers?"

Difference Between an Abstract and a Summary

Most reports consist of a synopsis, which is called an abstract or sometimes a summary. An abstract tells what the report is about and gives the extent of coverage. A summary gives the substance of the report without any illustrations and explanations. An abstract will give the method of analysis, the significant findings, the important conclusions and the major recommendations. The abstract is generally about two to three per cent of the original while the summary is about five to ten percent. In order to facilitate quick and easy comprehension for the reader, both these elements should be self-sufficient and intelligible, without reference to any other part of the report.

NOTE-TAKING

Note-taking is a skill widely used by students and teachers alike. It is primarily a way to filter and jot down the important ideas or comments. One can even make notes while going through an experiment. This would involve commenting on the problem, the outcome and probable solutions. Notes have to be taken in the classroom, while preparing for the examination and also while referring to reference material. Notes are also taken to jot down important points in a lecture, or to document an important discussion or presentation. These points can be put in a linear form where the main points and the sub-points are aligned differently to make the difference clear, or they can be put in the form of a flow chart where the sub points emerge from the nodal points. Finally, they can be even put in the circular form of a mind map where they branch out from the centre to the margins. (This form has been discussed in detail in the next chapter under mind-mapping.)

Given below are a few sentences and their main points. How can they be reduced to the form of notes?

1. a. Artificial intelligence is one of the most useful modules we have in this course
   b. Art intll—very useful module.
2. a. It is important to get into a pollution-free mode of production. Otherwise, the price we have to pay will be very heavy. Life under sea and on earth might even perish
   b. Pollution-free environment is a must. If not, life might end
      - On land
      - In the sea

An important part of note-taking is identifying the main and the sub-points in a piece of writing. It helps you organize the ideas, and comprehend the meanings as related, yet distinct chunks.

Look at the following example:

Recession

Economic growth occurs in a country when all the factories and offices are working and the national income is going up. History shows that in most countries economic growth is not constant. Even the advanced countries like the United States and Britain go into recession on a regular basis. Unemployment increases, factories and offices are closed down, and the economy produces fewer goods than it could if its resources were being fully utilized.

Sometimes the recession is particularly bad and develops into a depression. In the great depression of the 1930s, the national income of the United States fell from $95.2 billion in 1929 to $48.6 billion in 1933. One in four workers became unemployed. Reduced imports dragged a few other countries like Britain, Germany and Australia into depression too. Most people in a country are likely to suffer from the effects of a depression. Wages may fall, cutting the spending power of workers lucky enough to keep their jobs. The unemployed and their families are thrown into poverty. They have to rely on state benefits and charity for support.

One can make notes of this piece of writing in the following way:

1. Signs of economic growth
   - Working factories/offices
   - National income is up
2. Recession in advanced countries, a regular phenomenon
   - USA in 1930's
   - One in four became jobless
   - National income fell from $95.2 billion to $48.6 billion
   - Reduced imports that affected other countries too.
3. Impact of recession
   - Wages fall, cutting spending power.
   - Unemployment increases
   - Families, dependent on state support.

ACTIVITY

23. Given below are some paragraphs. Make notes; try to minimize them like the ones done before.
    a. In the countries of the western part of the world, particularly those of Europe, and America, rice (Oryza *sativa*) is not a very important cereal, yet, it is the chief food of about half of all the people in the world. This is because enormous numbers of people who live in China, Japan and India, as well as other parts of Asia such as Malaysia and the Philippine Islands, live mainly on rice. Rice yields more food per hectare than any other grain and Asia does not feed itself with any other crop. Although Asia produces more than nine-tenths of the world's output, rice has still to be imported in years of poor harvest. Time and again, in Asia, failure of the rice crop has resulted in famine and starvation.
    b. As the name implies, mini-computers are small computers. Unlike a mainframe computer which might fill an entire room, a mini-computer may fit in a single rack or box. In fact, it is a scaled down version of the main-frame computer.
    c. History is a process of recording the past. However, we should remember that past was not always recorded the way we do it today in history books. Writing on stone tablets, folklores, popular sayings, and temple documents were also recordings of history. But these mediums envisaged history differently. The present form of recording history, where authenticity and objectivity are emphasized, began only a century and half ago.
    d. The sun is the most direct source of energy. It powers the flow of wind and water cycle and sustains all life. Plants use this energy to synthesize carbohydrates from simple substances like carbon dioxide and water. All the food is derived from the process of photosynthesis. In fact, the energy by which animals including the human beings live is generated by oxidation of the food produced by the plants.

Points to Remember

- Make the notes brief:
  (Never use a sentence when you can use a phrase.
  Never use a phrase when you can use a word).
- Use abbreviations and symbols.
  (Remember to use them consistently).
- While note-taking, focus on outline and be aware of the main and the sub-points.

SUMMARY

- Correctness, precision, and elegance are very important for good writing.
- Careful word order is important for clear as well as unambiguous writing.

- Pre-writing, writing and post writing are the three important processes involved in any good writing.
- Paragraph along with sentence is the basic unit of writing. Proper sequencing is very important here.
- Précis writing is generally done to reduce the given paragraph to one third of its length.
- For summary writing, it is important to condense and also arrive at the most important points.
- Note-taking is a means to filter out important points and organize the paragraph into main and sub-points.

REVIEW QUESTIONS

1. What are the factors that contribute to clear, fluent, and effective writing?
2. What constitutes the content of any writing?
3. How do we organize our ideas while writing a paragraph?
4. What is a précis and what are the three main features of a good précis?
5. Discuss the notion of sequencing.
6. What do you understand by the term instructional writing?
7. What is the difference between abstract writing and summary writing?
8. What are the important features of note-making?

CHAPTER 7

Creativity and Mind-Mapping

Chapter Objectives

This chapter aims at examining the vital subjects of creativity and mind-mapping. It talks of the different manifestations of creativity and how they are important in our personal and academic life. The chapter aims to make readers aware of how creativity can be nurtured and used. It also discusses mind-mapping as one of the ways of expressing and developing creativity. It gives a detailed account of how mind-mapping can be seen as working with and complementing the functioning of the brain while being used effectively for organization and creativity simultaneously.

CREATIVITY

> *You have in you all and a thousand times more than is in all books.*
> *Never lose faith in yourself; you can do anything in this universe. Never weaken yourself, all power is yours.*
>
> —*Vivekananda*

Creativity can be described as a way of thinking. It is the manner of evolving new thoughts, approaching solutions differently, or simply adapting to a different manner of living or profession. With changing perceptions, we no longer believe that creativity is inborn. It is a potential that every one of us possesses. The difference lies in whether we have been able to nurture it, develop it, or tried to give it an expression and direction. Similarly, every situation has the potential to be creatively altered or improved. What is necessary is exploring alternatives, and finding ways of doing things.

TIMES WHEN WE ARE CREATIVE

Even the most creative people are not equally creative at all times. Similarly every person does become creative at certain times. Let us look at a few situations that are conducive for creativity, which make us think differently.

Thinking up a new idea

There are many occasions when suddenly we look at a problem differently, we find a new solution, and we phrase a thought or even try out a new recipe. All these, and many more are expressions of our creativity. At times, it can be the understanding of the source of disagreement or even a child discovering how to fix a set of puzzles. We have to understand that creativity isn't the realm of profound thinking only. Batting elegantly and with improvisation in the cricket field (or even writing a business report differently) is also creativity.

> Leonardo da Vinci, creator of Mona Lisa's smile was also the creator of the contact lens. Back in the 1500's, he found that the best way to correct poor vision was to put a short water-filled tube sealed at the end of the flat lens against the eye. This idea has become the contact lens of today.

> In 1904, Arnold Fornachou, an ice cream vendor exhausted all his paper dishes while serving icecreams to customers at a fair. Luckily, he was then standing opposite to a baker who was selling thin waffles. History has it that he bought a few, twisted them into cones, scooped ice creams into them and lo! the ice-cream cone was born.

Doing something spontaneous

Sometimes, in a moment of inspiration, we give a speech off-the-cuff, respond spontaneously. We may try a new route or decide to try out a new combination of clothes successfully–something we never thought would jell before. All these are outlets to our personal creativity. They are moments you cannot always repeat. These are also moments when we are at our best, competent and confident.

Creativity and building relationships

Fostering a meaningful relationship too means creativity. Sometimes we change our approach to meaningfully deal with a person. We steer a relationship to a win-win situation and create a context that can be based on collaboration and cooperation. The way we establish and nurture our relationships are indicative of the quality of our underlying creativity. The way we express ourselves in our relationships, too, manifests creativity.

I cannot teach anybody anything. I can only make them think.
—*Socrates*

A first-rate soup is more creative than a second-rate painting.
—*Abraham Maslow*

ACTIVITIES

1. Find out five ways to:
   a. avert environmental pollution 10 years from hence.
   b. make cooking easier.
   c. solve the problem of excess population 20 years hence.
   d. lower the stress level of your friends.
   e. convince your girl friend / boy friend to come to a movie with you.
2. List down the names of three people you dislike. However find 3 qualities in each of them, that is appreciable.
3. Make a note of two situations, which make you very uncomfortable. Devise 3 things you can do in each of the situations to be more at ease.

1. ____________________	2. ____________________
a.	a. ....
b.	b. ....
c.	c. ....

WAYS IN WHICH YOU CAN BE CREATIVE

Creativity manifests itself in many ways. Different people experience it differently. Some of the ways in which creativity blossoms are:

Creativity through Visioning

One way of finding creative solutions to problems is by visioning, through which you seek the ideal, long-term solution to problems. During visioning you let yourself go, imagine what it could be like if dreams came true or if problems were solved. It often helps in providing direction, inspiration and momentum to creative projects. Questions like: What would I like to be 10 years from hence? Help us form long-term objectives and focus our efforts."

> In 1961, J. F. Kennedy shared his vision with the nation. He had a dream that within the next ten years the U.S. would send a man to the moon. By the time Neil Armstrong finally took his "giant leap for mankind", Kennedy's vision had become the vision of the nation.

> In the 1850's a 17-year-old tailor noticed that cloth trousers of miners wore off very quickly. He then started stitching overalls made of a stiff canvass that could stand wear and tear. Years later he substituted the canvass with a fabric called denim and dyed it blue to minimize the soil stains. Thus came the blue jeans. The tailor's name was Levi Strauss.

Creativity through Exploring

This method of exploring your creativity involves questioning and challenging the core assumptions, making a radical breakthrough and coming up with new assumptions. This style of creativity works without any consideration of long-term goals, even consideration of the practical realities. The typical questions one asks in this process are: How can I break away from all the givens and establish an alternative?

> In 1948, a Swiss mountaineer was troubled by the prickly burrs that tightly clung on to his socks. As he sat tearing them off, an idea struck him. He studied the burr's clinging properties and invented a fastener that made the zipper obsolete. This was the invention of the velcro. The man's name was George De Maistred.

It is important to remember that each of our styles of creativity is a mixture of these. All of us function in the ways mentioned above at one time or the other.

ACTIVITY

4. Look at the questions given below. Categorize them into one of the types of creativity.
   a. What could give us a world-class processing unit?
   b. None of this works. How can we rewrite the rules of competition?
   c. How can we bring together the best we have in our different teams?
   d. How can we improve upon our core strengths?
   e. What should be our long-term goals and strategies.
   f. What could we possibly do to shake up things and bring about a total change?

DEVELOPING YOUR CREATIVITY

While talking about the ways in which we express our creativity and the styles we adopt to be creative, the constant question that comes up is how do we develop our creativity? Does it happen automatically or is there a way by which one can practice to make it happen.

If we analyze the nature of creativity we find in it a mixture of intuition and logical thinking. All discoveries have been made with logical thinking acting as the base for insight to flash and make the final connection or vice-versa. It is one followed by the other, or both operating together to form a cognizable whole. It is seen as a phenomenon of ideas coming from 'the back burner of our minds – cooking away' beyond our conscious awareness. Sometimes we even place it there unintentionally until it matures and pops into our minds'.

This can happen at a time when the mind is relatively rested–alert but not involved in a stream of thought. The art of developing creative skills, thus, is to train the mind to produce these "spaces", create opportunities for our intuition to act on our logic and form a whole. It is important, therefore, to create this intuitive space.

> Friedrich Kekule had been working for years to discover the molecular structure of benzene. One night, he says, he dozed off in front of the fireplace and saw a vision as if in a dream: "Again atoms were juggling before my eyes… everything was moving in a snakelike and twisted manner. Suddenly one of the snakes got hold of its own tail and the whole structure was mockingly twisted in front of my eyes. As if struck by a lightning, I awoke".
>
> He had envisioned the closed ring structure of benzene.
>
> Srinivasa Ramanujam, the Indian Mathemetician, is known to have reported that he arrived at the solutions to his problems in his dreams. He would dream of a stream of blood where goddess Kali would write out the solutions for him.

There is no logical path to the natural laws. Only intuition, resting on the sympathetic understanding of experience, can reach them.

—Einstein.

Creating the Intuitive Space

If we look into the history of inventions, we find that the most original ideas arise from the space in between our thoughts. To improve one's creativity, thus, it is important to understand these spaces. This is the phenomenon we can sometimes observe during anyone performing at his or her best. What is significant is the ease, the grace and the confidence that comes in during this state of consciousness. The act is done as if it were natural, as if there was no other way it could be done. We find it during an athlete's peak performance, a speaker giving one of the best addresses or a writer writing the most intense parts of his masterpiece – alert and active, yet loose and fluid.

Intuition is knowledge gained without rational thought. It comes from a stratum of awareness just below the conscious level; it is slippery and elusive. New ideas spring from a mind that organizes these experiences, facts and relationships to discern a path not taken before.

—Roy Rowan

Activity

5. Whenever you need to relax, close your eyes and visualize the following:
   a. The vast, expansive blue sea. Try to hear the roaring of the waves. Feel the waves dashing against your legs and retreating.
   b. Feel the cool evening breeze flowing over your skin. Try the recreate the cool, relaxed state it takes your mind into.
   c. Think of the hot blazing afternoon sun. Feel your skin scorching in the heat, your eyes glazed with the shine.
   d. Visualize a forest thick with trees. See the blanket of green over your head, hear the rustle of dry leaves beneath your feet.

FACTORS THAT BLOCK CREATIVITY

Look at the way a child plays. Carefully listen to the conversations they make. You will be amazed at the number of creative ideas that spring forth every second. Children are creative. But our adult life, our system of education and the thinking it promotes block our creativity, stops the flow. Many factors in our lives contribute to this. Some of them are:

Fear: Fear is one of the emotions that often forces us to limit ourselves. Fear of being rejected, fear of uncertainty, fear of failure and sometimes even fear of success constricts us and holds us back from moving ahead, finding new connections, exploring ourselves.

Anger and guilt: Guilt, depression or anger are emotions that blind us to the alternatives available to us. They prevent us from thinking of other possibilities, exploring all possible alternatives. At the workplace, they block us from fruitfully working together and bringing in positive vibrancy that can entirely alter the way we function and look at work.

Stress: Stress that comes from our urge for growth and expansion can be positive, but the stress that results from tension and anxiety can be damaging, can block the 'flow' in our thinking.

To realize your creative self, remember:

- Stress is often more psychological than real. If we enjoy the work we do, we rarely get stressed.
- Assume that all experiences can be positive and can help your personal growth.
- Envision what you want to be. Believe it, trust it.
- Allow yourself time. Innovative thinking has its own pace. Honour the seasons of creativity.

In the final analysis, it may be said that there is no one way to develop creativity. Every mind is unique. Every mind functions differently. So, the best that can be done is to observe and see. Find out what works best for you and your team and accordingly adopt the best way of working.

ACTIVITY

6. Think of some situations when you had to break the rules, go against people and aggressively implement your plans.

MIND-MAPPING: THE NETWORKING OF IDEAS

In the context of academic writing, one of the means of fostering creativity is by using mind-mapping. It can be seen as a technique to utilize our creativity to make our academic performance better. Mind-mapping can be seen as a graphic technique for representing ideas on a particular topic. Here, ideas move out from the center to the outer fringes, branching out and associating themselves with a network of other ideas. Every idea leads to a spray of associated ideas; each of these again, becomes the source of another spray of associations. This is a method primarily developed by Tony Buzan. His inquiry began from basic questions like

- How does one learn?
- What is the nature of thinking?
- What are the techniques of memorizing?
- What are the techniques of creative thinking?
- How can one read faster and more efficiently?

The Human Brain

The neural structure in the human brain is an amazing structure that looks almost like a super octopus. It is made up of branches and tentacles moving out of nodal points. Information passes through these branch-like structures called dendrites. This mechanism enables the cascading of bio-chemical information across the brain, from one cell to the other. A small brain cell can receive impulses from hundreds and thousands of connecting points every second like a vast sophisticated telephone exchange and redirect it along the appropriate path.

Every time you think a thought, the bio-chemical resistance along the way that carries the thought is reduced. The more you think a particular thought therefore, the easier it is for the brain to think the thought. A pattern is thus created and the brain re-creates the pattern. Learning puts it in the form of a familiar pattern, hence, is much more easily remembered and assimilated.

The Brain and the Gestalt

We have seen that the brain always creates and recreates patterns. Naturally, therefore, it always tends to look for patterns and completion. If we see five pieces of wood, laying close to one another our brain will invariably perceive a pattern and make a complete picture of disjointed wood pieces. Similarly, we perceive patterns in floating clouds and when someone begins a story and leaves it incomplete, our mind completes it and produces a well-designed whole. This is otherwise called the brain's capacity to perceive wholes and perceive in patterns. This, put very simply, is the mind's potential for Gestalt.

Radiant thinking

We have seen that our brain works through associations. Often a smell, a taste or a sight acts as a triggering factor and starts a chain of associations or connections. Every stimulus entering the brain is like a central sphere that radiates millions of hooks that have other associations connected with them. This process is otherwise called 'radiant thinking'– the inherent process of the brain's functioning where thoughts move or radiate from one or many given centers and proceed from one association to the next.

Mind map is the external manifestation of this internal process. It radiates or branches out from a center to different points and each point becomes, in turn, a sub-center of association. This can proceed infinitely, branching away or towards a common center.

The role of images and colours in remembering

The brain can be divided into two hemispheres—the left and the right. The left hemisphere is largely logical and it controls functions like the learning of words, logic, numbers, linearity, analysis and lists. The right hemisphere is more emotional. It helps in the perception of rhythm, pictures, imagination, colours and dimensions. In most people we have either the left or the right side of the brain more developed. Successful people often combine both for effective learning. They use the total brain – the logical as well as the imaginative faculty – evolve a learning strategy that draws from both and combines them effectively.

The most common mistake made when presenting material is that it is made monotonous with similar looking lists and lines. It hardly triggers the brain's creativity or does anything to aid association or memory. Simple tricks sometimes make learning and remembering much easier and pleasurable.

Colour variation or highlights of important points in different colors aids memory. Sometimes, different kinds of colour backgrounds are used to show the different aspects of the material being presented. Primarily, this helps the brain make and retain associations and note variations.

Using images for experiments have shown that the brain's capacity to store images is infinite. Any material linked to images, thus, has more chance of being retained. It breaks down the brain's resistance to learning and creates a pattern for the brain to rely on. Difficult words are thus better learnt if they can be broken down and each part associated with an image. It has also been seen that children who have been categorized as having less IQ learn better with images. (This also proves that our learning methods as well as IQ tests rely on only one kind of learning i.e., learning through words, numbers and logic). In all this, the fact that finally emerges is that imagination, pictures, colours and images are as important learning tools as words and numbers are. And one has to be able to creatively use these faculties for effective learning,

MIND-MAPPING AND THE LEARNING PROCESS

Mind-mapping, as we have seen, is an extremely powerful technique that can synthesize the various capacities of the brain and powerfully harmonize it to form a coherent whole, stimulate creative thinking. The spray of associations can radiate from various centers, and each can go on to become a center for another spray of associations. The associations can be documented through images, words, varied colours and even varying lines of thickness.

There is another advantage the mind-mapping technique has over the others. If you have forgotten a point or feel that there is more that can be added, just leave blanks. Your brain's natural tendency to fill up gaps will eventually supply the missing point. This is otherwise called the gestalt of the brain.

Some Guiding Principles in Making a Mind-Map

a. Use a central image.
b. Use images wherever possible
c. Use colours to separate one level of mental association from the other
d. Increase the line width to show main points and sub-points.
e. Underline, highlight, encircle or use any other means to highlight.
f. Use arrows or dotted lines when you want to make connections across the branches.
g. Use one key-word per line.
h. Write the key-words on lines.
i. Always ensure that the line length is equal to the word length.

Given in Figure 7.1 the picture of a typical mind-map.

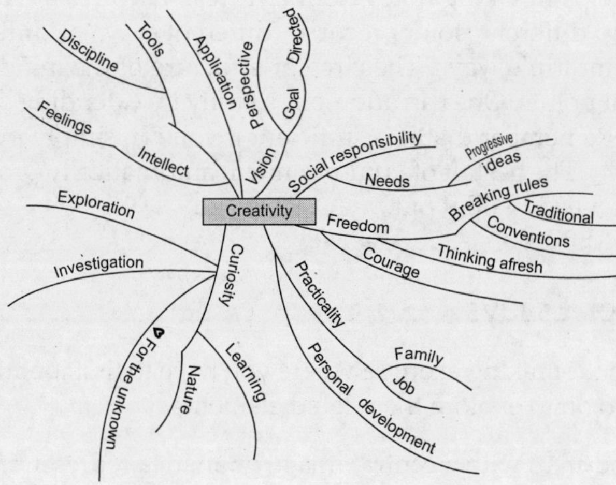

Figure 7.1 Mind-map of 'creativity'

Note-Taking and Mind-Mapping

Note-taking, as we have seen before, is the process of receiving other people's ideas from speeches or books and organizing them into a structure that reflects their thoughts. While note-taking, it is inevitable that note-takers' thoughts too become a part of the note. The points which he/she thinks are important, or the structure that she lends to the material that is to be transferred become very much a part of the final note that is made.

The linear pattern generally followed during note-taking does not allow one to make associations between points and sub-points, and they do not really take help from the association-making power of the brain. Also, it does not utilize the brains pattern-making capacity or its image-retaining ability.

Interestingly, if put in the form of a mind-map, the process of note-taking can be made creative and it can perform much more efficiently by taking help from the unutilized parts of the brain. The following steps can be taken to prepare a mind-map of the material you have for note-taking.

a. Very quickly browse through the entire material to have a general feel of how it is organized.
b. Mentally, make an overview of the major headings, results and conclusions. Also, think of the illustrations and graphics you might want to present. This will give a general idea of the central image, the main branches and sub-branches.
c. Now, it will be easy for you to fill out the mind-map. See where you would like to include the images, diagrams and establish the links. See if you would like to use codes and show the associated points. Use more than two colours to highlight the main-points. Use images wherever possible. It will trigger the association-making capacity in your brain.
d. Review the whole process again. Make sure that all the main points have been included. If you have left out any, complete it at a later stage.

Mind-mapping also helps in taking notes from lectures. Lectures are not always very structured. While lecturing, different points recur in different ways at different points and they also get associated in multiple ways. The circular structure of the mind map helps greatly in documenting the main points. One can add points easily by extending a branch and linking it up with other points. Remember that it is sufficient if you can simply put an image or a word condensing the thought. The gestalt of your brain will automatically provide the details of the information when you need to fill it out.

Mind-Mapping for Essays

An essay is a kind of note-making exercise where you have to first identify the essential topics of concern and then go on to explore the related elements.

a. Begin your mind-map with a central image or an idea representing the essay.
b. Decide on the basic ordering in terms of divisions and sub-divisions.
c. Allow your mind to move freely, documenting all the ideas that come. Keep in mind the basic ordering principles and the levels to which the ideas belong. At this stage, you'll need to be analytical, decide the hierarchy and arrangement of ideas.
d. Use different colours to show the different hierarchical structures. You could even use different thickness of lines for different levels. Use codes or symbols if you want certain points to be linked in the writing. Draw arrows to show the flow. Move back and forth from one point to the other. Let the whole of your brain have its full play.
e. If you face any problems in making associations or if you have a 'block', leave blank

lines and continue working. Your brain's natural capacity to form a whole will give you material and ideas to fill the gaps.

f. Review the map for any other information and missing links.
g. Begin the writing of the first draft of the essay. A well-ordered mind-map would give you an ordered structure of the main and sub-points. Study the links, consider the points and proceed with the main and sub-points.
h. At the end, review the essay while analyzing the mind-map. Give the finishing touches, ensure that you have sign posted the different points well. Check your use of linkers, and check for general coherence.
i. Reconsider your conclusion, review, modify and expand it, if necessary.

> The more I wrote and drew, the more things came to my mind. The more ideas I got, the more brave and original they were. I have realized that a mind map is never ending.
>
> —Kartarina, a student, while doing a mind map.

Mind-Mapping for Examinations

Answering the examination paper effectively has always remained a problem for students. Recollecting all the relevant points, putting them in order and then managing time – all these remain problems eternally for students to solve. Using mind-mapping during examinations can solve these problems to a large extent. Follow the steps given below to make effective mind-maps during examinations.

a. Read through the question paper thoroughly, choosing the questions you want to answer. Note down any point that occurs the your mind immediately.
b. Decide the order in which you want to answer them and the amount of time you want to spend on each of them.
c. Start making your mind-map. Note down all the relevant points that occur to you. Make the main points and sub-points related to each.
d. Give your brain full freedom to move to and fro, noting down all the points and associations that occur to you.
e. Try to organize, re-organize making references and cross-references.
f. Start elaborating your points—a paragraph or two for each of the branches. By now, you have all your main points as well as the sub-points. You should be able to assess how much time you will need to elaborate each.
g. Stick to your time limit and organize every answer systematically and well.

It is similarly helpful to prepare mind-maps before finally settling down to write a report or a project.

The gap between thinking and putting the words on paper is a process that is always troublesome and time-consuming. Mind-mapping is one technique that can do this efficiently. It allows one to sketch out the main ideas and quickly see how they relate to one another. It separates thinking from writing and when it is time to start writing, the ideas are already structured with clear focus and direction.

Very often in today's world, we stand at crossroads, confused over which career to choose, what decision to take, etc. Using mind-maps at such times can be of immense help. It can give you a better insight of yourself, your needs, desires and aims. It can also show you how you can analyze yourself. A mind-map well done can be a complete reflection of your personality. So quickly draw a colorful mind-map that captures your personality. It might give you vital clues about your strengths, weaknesses, your aptitude and dominant traits. The mind map format can also be very effectively used to conduct a brain storming session. Collate ideas and put them together.

Points to Remember

- Mind-mapping is a graphic technique for representing idea, using words, images, symbols and color.
- It is based on the pattern found in the natural architecture of the brain.
- It allows one to access multiple intelligences and generates new ideas.
- Since it allows us to quickly scan our mind, it can help formulate clear ideas about personal and professional goals.
- It can be an effective tool for note-making.
- Mind-mapping can also be helpful in writing reports, etc.

MIND-MAPPING: SOME DO'S AND DON'T'S

Do's

- Begin from the centre. Naturally the brain focuses on the center first, not the left hand top.
- Break stereotypes: Use butcher paper, colour pencils and pictures, however absurd.
- Think fast: Keep working
- Use symbols and key words. The brain works in bursts of 5–7 minutes. You should be able to capture the thoughts.
- Make the surroundings pleasant. It generates better ideas.
- Stand up and work. It energizes you.

Don't's

- Don't try to dismiss any idea, even if it appears irrelevant. Noting it down somewhere will help the brain relax and proceed.
- Don't stop very often to think. If nothing occurs, keep making empty lines. Your brains gestalt will fill it up later.
- Don't try very hard to organize when you are mapping. Note down all the ideas first; organization can come later.

ACTIVITIES

7. Given in Figure 7.2 is a half-completed mind map of the concept **home**. Complete it.

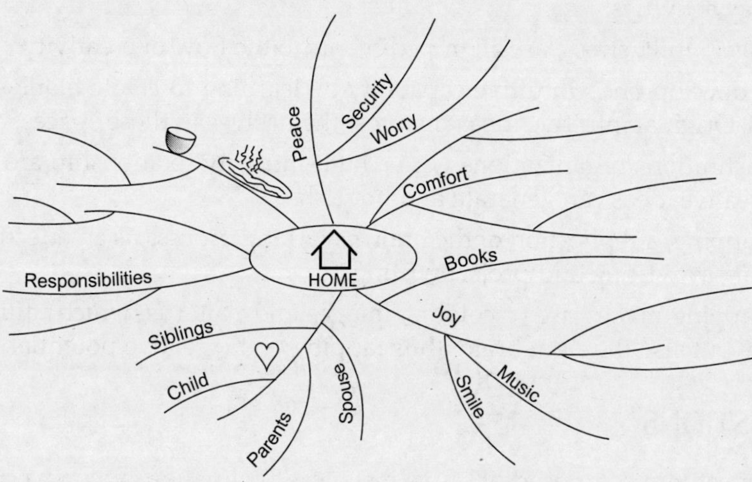

Figure 7.2 Mind-map of home

8. Add branches if you want to and compare it with what your friends have done. Use at least three colours, to show the hierarchy of ideas. Also, use codes or symbols to show associations among the branches.

9. Here is a write-up about craftsmanship. Prepare a mind-map of it. Use colors, codes, and symbols to make it striking.

Craftsmanship, then and now.

The advance of the age of machinery has not been all advantageous. In fact against all that the machine has given us, there is one serious disadvantage – the decline in craftsmanship. In days gone by, a furniture maker would use with care and pride the tools which, over a period of time, had become almost a part of him, and a chair took shape before his eyes. It was work not only of his hands but also of his mind, and expressed something of himself; no other chair, even one made by his own hands, would be similar. So it was with all craftsmen; everything they made was their own work, the result of their skill in the use of their tools, and they could look on it with pardonable pride.

What is the position today? In large factories of the machine age, rows of men are engaged in producing not a whole article, but merely one part of that article. The individual workman does not even have the satisfaction of feeling that this part is the work of his own hands, because it is made by a complicated machine. All he has to do is to feed the raw material into the machine, press a lever, and put the finished part on a moving belt, which will convey it to the assembly lines. Thus the modern worker is denied the joy of creation; his work is reduced to a monotonous repetition of automatic movements that are carried out almost without the use of his conscious mind.

SUMMARY

- Creativity today has been redefined as the capacity to grow to one's fullest potential, and survive in a highly competitive environment or even efficiently function in one's day-to-day activities.
- Fears, anger, guilt, stress, are all impediments to the flow of creativity.
- One can develop one's intuitive capability by learning to create blank spaces between thoughts. Original intuitive ideas manifest themselves in these spaces.
- Teams/institutions/organizations with a high-integrity relationship are fertile grounds where creative ideas can generate and flourish.
- Mind-mapping, a replication of the brain's working process, is an effective tool for note making, essay writing and project writing.
- Mind-mapping makes use of colours, images and patterns, which utilize not only the left brain, but also the right brain, thus tapping one's creative potential.

REVIEW QUESTIONS

1. What is creativity composed of? Give instances of situations when we are creative.
2. Does creativity play an important role in determining our relationships around us? Explain.
3. Can one develop creativity? Expand.
4. What are the factors that block creativity?
5. The mind-mapping technique closely resembles the brain's information processing system. Explain.
6. Do you think that the traditional ways of teaching and learning do not compliment the brains natural learning or information-retaining system? Giver reasons for your answer.
7. What are the functions of the right brain and the left brain? State how the technique of mind-mapping can use both effectively and make different types of academic-writing better.

CHAPTER 8

Résumé Writing, Curriculum Vitae (CV) and Statement of Purpose (SOP)

Chapter Objectives

This chapter discusses the importance of writing effectively the résumé, the CV and the statement of purpose (SOP). This is discussed in the context of their high career making value in relation to getting job calls and academic admissions.

DEFINITION OF A RÉSUMÉ

A résumé is a one or two page summary of your education, skills, accomplishments, and experience. Your résumé's purpose is to enable you to gain the selector's attention. A résumé does its job successfully if it does not exclude you from their consideration. To prepare a successful résumé, you need to know how to review, summarize, and present your experiences and achievements on one page. Unless you have considerable experience, you don't need two pages. Outline your achievements briefly and concisely. Your résumé is your ticket to an interview where you can sell yourself!

Writing a résumé is serious business. Often, it is the first impression you will make on a prospective employer. Hopefully, after looking over your résumé, the employer will grant you the opportunity to make a second impression. If we look at job search as a marketing campaign, we can then look at the résumé as a print advertisement or a marketing brochure. If you take a look through a magazine you will see many advertisements. Try to find one that tells you to buy a product because the company needs to increase its profits. You will be hard-pressed to find such a insensitive one.

The advertisements you see tell you what the manufacturer's product can do for you – make your smile bright, your hair shiny, or simply make your life better. When putting together your résumé, evaluate the needs of the employer and then determine how you can fill those needs. If you have access to a computer and a quality printer, you can design a targeted résumé for every job that you apply for. If you have to mass produce your résumé, you will have to do a little guesswork to come up with one that will impress everyone. There are three types of résumés. They are, (a) functional résumé, (b) chronological résumé and (c) combined résumé.

Functional Résumé

The functional résumé focuses on your skills and accomplishments. It highlights what they are, not when you developed them. It is best for:

a. People with lots of job experience and many jobs.
b. People just entering the work force with no track record.
c. People who are returning to work after a long absence.
d. People who are changing careers who want to highlight their skills and credentials.
e. People who are closer to retirement age.
f. People whose career growth has not been good.
g. Military personnel who are seeking civilian jobs.

Its contents include:

a. Contact information
b. Objective
c. Skills
d. Work experience
e. Education

Chronological Résumé

The chronological résumé is the most common. It is a chronological listing of your jobs and experience with most recent mentioned first. It is best for:

a. People who have practical work experience without long periods of employment and minimal job changes.
b. People who have demonstrated growth in a single profession.

Its contents include:

a. Contact information
b. Objective
c. Work experience
d. Education

Combined Résumé

A combined résumé includes elements of both the chronological and functional formats. It may be a shorter chronology of job descriptions preceded by a short 'Skills and Accomplishments' section (or with a longer summary including a skills list or a list of 'qualifications'); or, it may be a standard functional résumé with the accomplishments under headings of different jobs held.

There are obvious advantages to this combined approach: It maximizes the advantages of both kinds of résumés, avoiding potential negative effects of either type. One disadvantage is that it tends to be a longer résumé. Another is that it can be repetitious: Accomplishments and skills may have to be repeated in both the 'functional' section and the 'chronological' job descriptions

Power Words

Use power words to describe your experience and accomplishments. Here are some power words to use:

accelerated, accomplished, achieved, addressed, administered, advised, allocated, answered, appeared, applied, appointed, appraised, approved, arranged, assessed, assigned, assisted, assumed, assured, audited, awarded, ability

bought, briefed, broadened, brought, budgeted, built

capable, capability, cataloged, caused, changed, chaired, clarified, classified, closed, collected, combined, commented, communicated, compared, compiled, completed, computed, conceived, concluded, conducted, conceptualized, considered, consolidated, constructed, consulted, continued, contracted, controlled, converted, coordinated, corrected, counseled, counted, created, critiqued, cut, capacity, competence, competent, complete, completely, consistent, contributions

dealt, decided, defined, delegated, delivered, demonstrated, described, designed, determined, developed, devised, diagnosed, directed, discussed, distributed, documented, doubled, drafted, demonstrated, developing

earned, edited, effected, eliminated, endorsed, enlarged, enlisted, ensured, entered, established, estimated, evaluated, examined, executed, expanded, expedited, experienced, experimented, explained, explored, expressed, extended, educated, efficient, effective, effectiveness, enlarging, equipped, excellent, exceptional, expanding, experienced

filed, filled, financed, focused, forecast, formulated, found, founded

gathered, generated, graded, granted, guided, global

halved, handled, helped

identified, implemented, improved, incorporated, increased, indexed, initiated, influenced, innovated, inspected, installed, instituted, instructed, insured, interpreted, interviewed, introduced, invented, invested, investigated, involved, issued, increasing

joined

kept, knowledgeable

launched, learned, leased, lectured, led, licensed, listed, logged made, maintained, managed, matched, measured, mediated, met, modified, monitored, motivated, moved, major, mature, maturity,

named, navigated, negotiated, nationwide

observed, opened, operated, ordered, organized, oversaw, outstanding,

participated, perceived, performed, persuaded, planned, prepared, presented, processed, procured, programmed, prohibited, projected, promoted, proposed, provided, published, purchased, pursued, performance, positive, potential, productive, proficient, profitable, proven,

qualified, questioned, ,

raised, ranked, rated, realized, received, recommended, reconciled, recorded, recruited, redesigned, reduced, regulated, rehabilitated, related, reorganized, repaired, replaced, replied, reported, represented, researched, resolved, responded, restored, revamped, reviewed, revise, record, repeatedly, resourceful, responsible, results

saved, scheduled, selected, served, serviced, set, set, up, shaped, shared, showed, simplified, sold, solved, sorted, sought, sparked, specified, spoke, staffed, started, streamlined, strengthened, stressed, stretched, structured, studied, submitted, substituted, succeeded, suggested, summarized, superseded, supervised, surveyed, systematized, significant, significantly, sound, specialist, substantial, substantially, successful, stable,

tackled, targeted, taught, terminated, tested, took, toured, traced, tracked, traded, trained, transferred, transcribed, transformed, translated, transported, traveled, treated, trimmed, tripled, turned, tutored, thorough, thoroughly,

umpired, uncovered, understood, understudied, unified, unraveled, updated, upgraded, used, utilized

verbalized, verified, versatile, vigourous, visited

waged, weighed, widened, won, worked, wrote, well educated, well rounded, worldwide

GUIDELINES FOR EFFECTIVE RÉSUMÉ WRITING

- Be clear, direct, effective and professional. Make your résumé easy to read.
- Use bulleted statements to facilitate skimming. Avoid long paragraphs and large blocks of text.
- Try to keep it to one or two pages. If you have less than 10 years of experience, a single-page résumé is a good rule-of-thumb. The reader should be able to skim your résumé in 25 – 30 seconds.
- Determine a format and stick with it through the résumé i.e., ensure section headings have the same placement and font treatment throughout the résumé.
- Put dates next to the right-hand margins.
- Information about your most recent job may use current or past tense depending on your current status, only use past tense to describe previous accomplishments, since they are completed actions.

- Accurately use up-to-date terminology relevant to the industry you have targeted.
- Spell out terms. Avoid abbreviations and acronyms.
- Identifying information should be listed on the firstpage in a balanced, organized layout, including:
  - Name (should stand out, i.e., bold, all capital letters etc.)
  - Street address
  - City / state / zip code
  - Home, Mobile and/or Cell Phone (include 10-digit telephone numbers with area code)
  - Work phone
  - Email address (include personal not work)
- If your résumé is more than a single page, be sure to place name and page number in upper right hand corner of the second page.

Points to be Included

a. *Name and gender*
   Given the first and last name only because that's how you are introduced and introduce yourself. Sometimes mentioning the gender is useful especially where names are not revelatory of it in themselves and when you are sure you will not be discriminated because of your gender.

b. *Address*
   Give complete address. Do not abbreviate.

c. *Telephone number with area code.*

d. *Objective*
   Keep it short in just one or two sentences. Adjust to fit the position for which you are applying. If résumé is going to be circulated a lot, the broader the better.

e. *Education*
   If you are a recent graduate, place education before work experience, if not, place after.

f. *Skills*
   List all that are appropriate to the type of work you are seeking. Include computer skills and languages (understood, read, or spoken) for every job.

g. *Dates*
   Include some type of reference of when you had the job. Be consistent with your notation

h. *Job titles*
   Adjust to fit the position for which you are applying. If résumé is going to be circulated a lot, the broader the better. If appropriate, use the title of the position for which you are applying.

i. *Company name*
The company with whom you were employed. City and state are sufficient for the address.

j. *Responsibilities*
This is an essential part of the résumé. Highlight the responsibilities of your previous jobs that are related to the position for which you are applying.

k. *Professional licenses*
Include if important to line of work

l. *Publications and patents*
Use if important to your field or profession.

m. *Professional affiliations*
Exclude mention of political, social, religious or any other controvesial group. The emphasis is on your professional memberships, not personal.

Points to be generally excluded

a. *A résumé title*
It should be obvious what it is.

b. *Availability*
Apparently you are available; you are looking for work. It decreases the life span of your résumé and your efficiency if you do not get a job by the specified date.

c. *Salary*
If your request is too high, you are eliminated immediately. If it's too low, they may still trash your résumé, or worse they may pay you what you asked, which is thousands less than you are worth.

d. *Mention of age, race, religion, or national origin*
It is just not good business sense to mention these. Discrimination does happen to everyone now and then.

e. *Photographs*
Unnecessary, unless applying for a modeling or acting career. Then, a portfolio is recommended.

f. *Charts and graphs*
Nobody's résumé should have that much void space. You can do better.

g. *Weaknesses*
It is counter-productive. The purpose of the résumé is to accentuate the positives.

h. *Reason for leaving*
It is inappropriate for the résumé. If the employer wants to know, he or she will ask you.

i. *References*
Do not list references because it is unprofessional. State instead, "References are available upon request," at the very end of your résumé.

A Sample Résumé

a. **Name and Gender:**
GYAN PRAKASH (Male)

b. **Address:**
H.No: 1149, B.Nagar, Secunderabad, AP, India Pin: 500061

c. **Telephone:**
091 40 2707 0798, Cell: 98482 49777

d. **Objective:**
A software sales position in an organization, seeking an extraordinary record of generating new accounts, exceeding sales targets and enthusiastic customer relations.

e. **Education:**
Bachelor of Engineering (Computer);
Master of Business Administration (MBA-Marketing)

f. **Skills**
Computer skills (hardware and software)

Languages:

Understood:	English, Telugu, Oriya, Tamil, Kannada, Urdu, Hindi, Bengali, French
Read:	English, Telugu, and Hindi
Spoken:	English, Telugu, and Hindi
Interpersonal skills:	Wide network of friends and contacts.

g. **Job Titles:**
No work experience/ fresh graduate/ seeking job for the first time.

h. **Professional Affiliations**
Member of State Management Graduates Association

CURRICULUM VITAE (CV)

A curriculum vitae more commonly referred to as CV, is a longer (two or more pages) and a more detailed synopsis than a résumé. It comprises a summary of your educational and academic background, as well as teaching and research experience, publications, presentations, awards, honors, affiliations, and other details. Your CV ought to be clear, concise, complete, and up-to-date with current employment and educational information. The following is an appropriate format of globally acceptable curriculum vitae.

Curriculum Vitae Format I

Your Contact Information
Name, address, telephone, cell phone, email.

Personal Information
Date of birth, place of birth, citizenship, passport no / visa status, gender

Optional Personal Information
Marital status, spouse's name, children

Education:

Employment History

List in chronological order, include position details and dates work history, Academic Positions

Research Skills:

Presentations:

Publications:

Grants and Fellowships:

Awards and Honors

Skills and Qualifications:

References:

Curriculum Vitae Format 2

Name

Date of Birth:

Citizenship:

Address:

Phone:

Off:

Phone: Fax:

E-Mail:

Academic Degrees Earned:

Academic Positions Held:

Research Interests:

Publications:

Teaching Experience:

Conferences Attended:

Future Invitations:

Refereeing and Book Reviewing:

Seminar and Conference Organization:

Prizes and Awards:

Computer Skills:

References:

Sample Curriculum Vitae I

Sunil Kumar
Street, City, State, Zip
Phone: 888-777-2345
Cell: 255-366-466
email@email.com

Personal Information
Date of Birth: 17-08-1968
Place of Birth: Delhi
Citizenship: Indian
Passport no / Visa Status:
Gender: Male

Optional Personal Information
Marital Status: Married
Spouse's Name: Anita
Children: 2

Education:
Ph.D., Psychology, University of Hyderabad, 2006
Thrust areas: Psychology, Community Psychology
Dissertation: A Study of Learning Disabled Children from Low Income Groups

M.A., Psychology, M.K.G University at Visakhapatnam, 2003
Optionals: Psychology, Special Education

Thesis: Speaking Skills of Learning Disabled Children
B.A, Psychology, Delhi State University, New Delhi, 2000

Employment History
Academic Positions: Associate Professor, Psychology

Instructor, 2004 – 2006
University of Hyderabad
Course: General Psychology

Teaching Assistant, 2002 - 2003
M.K.G University
Courses: Special Education, Learning Disabilities

Research Skills:
Extensive knowledge of statistical programs.

Presentations:
Sunil Kumar(2006)., The behaviour of learning disabled adults in the classrooms. Paper presented at the Psychology Conference at the University of Hyderabad.

Publications:
Sunil Kumar (2005)., The behaviour of learning disabled adults in the classroom. *Journal of Educational Psychology*, 12 – 15.

Grants and Fellowships:
Sigmund Grant (University of Hyderabad Research Grant, 2005), Rs.48,000
Workshop Grant (for ASPA meeting in Mumbai, 2004), Rs.5,000

Awards and Honors:
Jung Scholar, 2005
Academic Excellence Award, 2003

Skills and Qualifications:
Microsoft Office, Internet
Programming ability in C++ and PHP
Fluent in Hindi, English, and French

References:
Prof. A. K. Mehrotra, Dept. of Psychology, University of Hyderabad, AP, India
Prof. A. M. Rao, Dept. of Philosophy, Delhi State University, New Delhi, India

Your CV is your own logo and also the company profile of you as a brand!

Are you able to market your functional skills, expertise and achievements? Is your résumé a mere reflection of your skills? If you are looking for a job, then it is important that you understand how best to offer yourself to an employer. This is done by writing a 'CV' (curriculum vitae – Latin for 'life story'), called 'Bio Data' in some countries.

Companies receive hundreds of résumés/CVs for one job opening. However, only few are short-listed for the interview. This short listing is done on the basis of your résumé/CV. On an average a Curriculum Vitae will receive no more than 30 seconds of initial consideration

In that time, it needs to make an impact. To get someone to look at it longer, it must quickly convey your capabilities, competence and 'essence'.

We need to understand the difference between the scope of the résumé/ CV of a thorough professional and that of a new recruit. With years of experience behind you, your résumé should focus on your strategic capabilities and your business acumen. It should document how your presence had been valuable in driving the organization up the ladder of corporate success.

When it comes to salary negotiations, a well-written CV plays an important role. If your CV conveys your full worth you are likely to get a higher salary offer than you might have done with a poorer CV. A CV is personal, but it is neither a biography, nor a paragraph, nor even a novel, in which you tell the story of your professional life!

Sample Curriculum Vitae II

Kalikrishna
228 SMR Layout, Hosur Rd, Bangalore 560012, India
Email: kalikrish_08@userhome.com
Phone: 91-80-2345-5678

EDUCATION

B.Tech. Computer Science and Engineering, Indian Institute of Technology, Mumbai (2002 – 2006)

Thesis Title: Effective Algorithm for Terrain Simplification – Fast Rendering
Advisor: Subhadra Rao

Summary: Improves the state of the art in occlusion plane detection, given terrain data. My implementation showed .... 2 – 3 lines

St. Xavier's School, Mumbai (Graduated 2000)

Ranked 1st in school in 12th C.B.S.E. Board Examination.

PROJECT WORK (B. Tech)

- Developed an optimizing compiler for mC++, a C++ subset with support for dynamic object migration over the network between compatible type-spaces.
- Designed a user-level distributed file system based on NFS with write-through caching, fault tolerance and consistency guarantees.

WORK EXPERIENCE

- Research Assistant, TIFR, Mumbai (Aug – Dec 200X): implemented a library of image processing functions for edge detection and de-skewing on scanned images. Adapted an off-the-shelf OCR package to operate on scanned mail images with 95% address recognition accuracy at the city/pin-code level and 90% at the street level.

- Project Trainee, Spontaneous Networks, Bangalore: (Jan 2005-present): Joined a 17-person startup implementing campus-wide video-on-demand system for corporate training. Implemented the streaming video component with buffering for jitter reduction. Also, bit-rate reduction in the event of congestion to meet frame-rate guarantees. Extensive performance testing was conducted.

COMPUTER SKILLS

- Software: SQL Server, Apache, CVS, Mathematica, Latex. Also, audio/video formats and codecs. Languages: C/C++, STL,
- Platforms: Linux, FreeBSD, NT 4.0, 2000.

PUBLICATIONS

Occlusion Culling using Hyperplane Projection, Anil Choudhary, and Subhadra Rao. Graphics Interface '01(2004) pp. 323–333.

AWARDS

- Best B.Tech thesis – 2004. Dept. of Computer Science, IIT Mumbai; Medalist at the International Mathematical Olympiad (IMO), 2003, Seoul, S. Korea.
- Ranked 26th in IIT Joint Entrance Examination – 2001.

REFERENCES

- Prof. Aloknath Kumar, Dept. of Computer Science, IIT Mumbai; Prof. Subhadra Rao, Dept. of Computer Science, IIT Mumbai
- Dr. G. V. K Raju, TIFR, Mumbai

STATEMENT OF PURPOSE (SOP)

It is an essay written by a prospective candidate for admissions into any academic programme.

How to write an effective statement of purpose?

1. **Establish your purpose in writing the statement**

   Usually the purpose is to persuade a particular admissions committee that you are an applicant they should choose. You may want to show that you have the ability and motivation to succeed in your field, or you may want to show that, on the basis of your experience, you are the kind of candidate who will do well in the field. Whatever the purpose, it must be explicit to give coherence to the whole statement.

   a. Pay attention to the purpose throughout the statement so that irrelevant material is excluded.

   b. Keep in mind the audience (committee) throughout the statement. Your audience is made up of faculty members who are experts in their field. They want to know that you can effectively go through the paces of learning, given your purpose.

2. **Establish the content of your statement**

   Sometimes, there are pre-given questions in the admission document itself. Ensure that you answer such direct questions completely. Analyze the questions or guidance statements thoroughly and answer all parts to build the statement.

   For instance: "What are the strengths and weaknesses in setting and achieving goals and working *through* people?" (Observe closely the small words. This example question says through people not with people, if it says *with* people, answer that way). This question points at six parts to be answered. They are,

   a. Strengths in setting goals,
   b. Strengths in achieving goals,
   c. Strengths in working through people,
   d. Weaknesses in setting goals,
   e. Weaknesses in achieving goals and
   f. weaknesses in working through people.

Usually graduate and professional schools are interested in the following:

a. *Your purpose in graduate study*. This implies that you must have thought this through before you try to answer the question.
b. *The area of study in which you wish to specialize*. This demands that you know the field well enough to make such decision.
c. *Your future use of your graduate study*. This would comprise career goals and plans for your future.
d. *Your special preparation and fitness for study in the field*. This is the context to link your academic background with your extracurricular experience to show how they combine to make you an exceptional candidate.
e. *Any problems or inconsistencies in your records or scores* such as a bad semester. Be confident to explain positively and justify the explanation. Since this is a refusal argument, it should be followed by a positive statement about your skills and talents.
f. *Any special conditions that are not revealed elsewhere* in the application such as a substantial (25 hour a week) work load outside of school. This also should be followed with an affirmative statement about yourself and your future.
g. You may be asked, "Why do you wish to attend this school?" This demands that you have obtained prior information about the school and realize its special appeal to you.
h. Most importantly, the statement of purpose ought to comprise information about you as a person. They know nothing about you that you don't tell them. *You* are the subject of the statement.

3. **Determine your approach and the style of the statement**

   There is no such thing as "the perfect way to write a statement." There is only the one that is best for you and fits your circumstances.

1. There are some things the statement should not be:
   a. Avoid the "what I did with my life" approach. This would be fine for school essays on "what I did last vacation." It is not good for a personal statement.

b. Equally simple is the approach "I've always wanted to be a __________." This is only suitable if it also reflects your current career goals.
c. Also avoid a statement that indicates your or your family member's ill-health.

These are some things the statement should do:

a. It should be objective while being self-revelatory. Write directly and in a straightforward manner that tells about your experience and what it means to you.
b. It should form conclusions that explain the value and meaning of your experiences such as: (a) what you learned about yourself; (b) about your field; (c) about your future goals; and (d) about your career concerns.
c. It should be specific. Document your conclusions with specific examples or draw your conclusions as the result of individual experience..
d. It should be an example of careful, credible writing.

Considerations about Form and Language

1. Keep to the page limit. Reviewers have to read many of these applications; don't overstrain them with extra pages.
2. Do not leave any typographical errors and do not use unacceptable and inappropriate language. You should not want to be taken less seriously due to these.

Words and phrases to be used discretely and with explanation

Significant, Invaluable, appealing to me, interesting, exciting, excited, appealing aspect, challenging, enjoyable, enjoy, I like it, satisfying, satisfaction, I can contribute, it's important, rewarding, valuable, fascinating , gratifying, helpful, appreciate, meaningful, useful, helping people, meant a lot to me, feel good, I like to help, stimulating, remarkable, people, incredible

Sample SOP I

Graduate School/ Engineering and Computer Science MS 1

I have been interested in problem solving from a very young age, especially problems relating to mathematics. The computer also has been part of my home for many years and I feel most comfortable working on it. My undergraduate studies have included Software Engineering, Computer Graphics and Visualization, Human Computer Interaction, Design and Analysis of Algorithms, Computational Mathematics, and Formal Logic for use with artificial intelligence systems. My own sphere of special interest in the field of computer science comprises the dynamic relationship between software and intelligence systems.

Now that I have completed my bachelor's degree, I look forward to a new challenge of studying in a graduate school. I see graduate school in Computer

Science as a new level of growth where I can reach an advanced level of professional ability that I shall seek to refine during the later part of my life. To end this, I am applying to study towards an M.S. degree in Engineering at the Department of Computer Science at the University of XXXX because I feel very strongly that this is the best location for continuing my studies in the area that I am most interested. I look forward to direct access to the Department of Mathematics Statistical Consulting Service and its assistance with experimental design, data display and analysis, and the interpretation of findings relating to statistical software engineering.

I would like to study in XXXX as your program ranks among the very best. I strongly believe that I am an ideal candidate for your program because I am an extremely dedicated worker who never ceases to explore most cutting-edge technologies. I have very strong analytical and problem-solving skills and I am a team player who looks forward to working with development teams in the advancement of software procedures, methods, tests, and systems. Furthermore, I am keen to participate in advanced testing, planning, and scheduling in order to assure that software products are optimally efficient and fulfill the purposes for which they were developed. Towards this, I have labored with algorithms and have mastered the basics of Formal Logic, always with an eye on artificial intelligence.

I especially look forward to working with problems and issues that have to deal with cognitive sciences, neural networks, and the development of AI solutions and products that will be useful for software of the future. Moreover, I am interested in computer graphics, visualization and am efficient in the implementation of 2D and 3D modeling using OpenGL. I would like to someday design and develop algorithms and new statistical methods for software engineering that would better integrate computer graphics into software visualization, thus exploiting the advancement of computer graphic processors that are now able to handle complex algorithms, for visual development. In this context, I would like to inform you that I was chosen to be a member of the International XXX Society and that XXXX, Inc had contacted me regarding a position with their company as an engineering intern. I had completed various projects including the development of underlying email support (90%), X, Y, Z etc.

I would like to thank you for having considered my application.

Sample SOP II

Graduate School/Mechanical Engineering MS 1

I am from India who determinedly wants to study the M.S. Degree program in Mechanical Engineering at your XXXX University. Last year, I completed my undergraduate degree in XX in India with a special focus in the area of production. I am therefore keen on doing graduate work in the area of Manufacturing

Engineering, Mechanical Engineering, and Operational Research. I bring in lot of dedication in my studies and my long-term goals include studying towards a Ph.D. degree after completing my M.S. For the future, I am planning to study Game Theory since I have a fascination for this area.

During my undergraduate studies, I was a member of the Production Engineering Student Association, and in this capacity I organized various paper presentations, computer games, and seminars. This experience was especially helpful for learning how to work with a team and I think that these experiences will be helpful to me at the graduate school, especially when it comes to working with a diverse group of people from all over the world. Furthermore, I have also other social interests and am very determined to give something back to society. This is why I am an active member of the National Service Scheme, have participated in a Blood Donation Camp and also worked on tree plantation work, in addition to other community services.

I am currently working for XXXX Ltd as a Graduate Engineer Trainee. This is India's largest diesel engine manufacturing company. This forthcoming July, I will have one year experience in diesel manufacturing and I am sure that this will also be helpful to me in doing well at the graduate school. I worked hard to develop my leadership and team skills throughout my college days and am presently working with one of the biggest manufacturing companies in India, I am learning the art of negotiation, and how to efficiently manage people. I feel that the greatest contribution that I would be able to make to society in the future would be to distinguish myself in the area of Operations Research.

Considering all of this, I earnestly feel that I shall do well at the Graduate school and I would like to thank you for having considered my application.

SUMMARY

- A résumé is a one or two page summary of your education, skills, accomplishments, and experience.
- To prepare a successful résumé, you need to know how to review, summarize, and present your experiences and achievements on one page.
- There are three types of résumés. They are functional résumé, chronological résumé, and combined résumé
- Use appropriate power words to describe your experience and accomplishments
- A curriculum vitae (CV), is a longer and a more detailed synopsis than a résumé.
- It comprises a summary of your educational and academic background, as well as teaching and research experience, publications, presentations, awards, honors, affiliations, and other details.
- Your CV ought to be clear, concise, complete, and up-to-date with current employment and educational information.
- Statement of purpose (SOP) is an essay written by a prospective candidate for admissions into any academic programme.
- Usually the purpose of an SOP is to persuade a particular admissions committee that you are an applicant they should choose.

REVIEW QUESTIONS

1. What is a résumé?
2. What are the features of an effective résumé?
3. What are the aspects to be avoided while writing a résumé?
4. What is a CV and what does it consist of?
5. How does a well-written CV help in negotiating the salary?
6. Recalling and analyzing experience – write short paragraphs on the following:
   a. Pick a memorable accomplishment in your life. What did you do? How did you accomplish it?
   b. What sort of important activities have you engaged in? With whom? What role did you play?
   c. What work experiences have you had? What was your job? Responsibility? How did you carry it out?
7. Write two short paragraphs on the following:
   a. What career have you chosen? What factors made you take this decision?
   b. What evidence shows that this is a correct choice? That is, how can you show that this choice is realistic? (Personal experience in the field is a good place to begin.)

CHAPTER 9

Team-Talk, Group Discussion and Interviews

Chapter Objectives

This chapter discusses the importance of talking effectively among members of a team and the dynamics of such a talk including conflict management. It spells out different team types. It distinguishes and explains the different roles which members of a team take on at different times. The chapter also deals extensively with the dynamics of group discussion and interview.

> *Coming together is a beginning, staying together is progress, and working together is success.*
>
> —*Henry Ford*

IMPORTANCE OF TALK IN A TEAM

Teamwork, as we all know, has become a very important part of corporate work culture today. Forming the right kind of team, causing the right chemistry to work and creating the right working atmosphere form an important part of getting the right results. It is very important, thus, to have a proper understanding of a team. One effective way of recognizing a well-struck team is to understand team-talk. Teams work together through conversations, whatever the channel might be – phone, fax, mail or face-to-face.

Language plays a very important role in teamwork. It reflects and even creates thoughts and feelings in both the speaker and the listener. It enhances or diminishes relationships, works towards learning and problem-solving in a team. Talk is also an indicator of how the team relates itself to the organization. It provides a perspective of the team dynamics – how the team members think, feel and act with one another. Look at the following statements:

- I am a team player, but I want rewards to be individual.
- I feel my image suffers by their lack of performance.
- The team leaders here are not experienced in managing conflict.
- I have the primary responsibility for the product, but I'm also dependent on them to achieve my own objectives.

All these are statements often heard and expressed during teamwork. They show the inevitability of individual initiative and responsibility along with the necessity of teamwork. They also reveal a paradox between independence and interdependence, a primary contradiction that all teams have inbuilt in their structure. Team members necessarily have difference in terms of knowledge, skills and experience. Without these differences, in fact, it is difficult to manage a team task. At the same time, it is also necessary that these differential factors are combined and integrated as one.

The paradox in a team is the contradiction between differentiation and integration. Similar is the case with identity and interdependence. Every member enters the team as a specialist with a distinct identity. But at the same time, a group can function well only if the members can effectively depend on one another. For group work, individual and independent judgement is as important as group thinking. The third paradox every team operates with is the paradox of trust. Team members have to trust one another while at the same time, remain vigilant. There has to be the right balance between trust and scrutiny. While trust, in this context, is very important for the collective functioning of the team, a constant testing and assessment of others' opinion is also important.

> *Many of us are more capable than some of us . . . but none of us is as capable as all of us!*
>
> *—Tom Wilson*

Activity

1. Look at the following statements and discuss the kinds of paradox they reveal.
    a. You, as an individual, succeed in achieving your goal. However, you are expected to show a willingness to set aside personal gain for the good of the organization.
    b. There is a confusion between her and me. There is a crossover in tasks. She is straight an R & D person. I'm from the manufacturing division. However, I shall now be doing product development work and she will be doing manufacturing.
    c. You need a level of trust in teams. Unless there is that mutual trust, people won't let their guard down. And unless they come out, they cannot perform.
    d. You have to be able to let go of your own ego. You have to give up your own ideas in favour of the team consensus. This is tough.
    e. The power of a good team is to convert differences into diversity that enriches itself as a whole.

f. As an engineering person, I try to meet my deadlines relentlessly, but as a member of this team, I have to take the other circumstances into consideration.
g. When I feel frustrated, it feels like there are two extremes. The best thing to do is to be nice to people. But then I would get the axe because the management sees me as not being able to deliver.
h. I don't want to tell them they can take more time, because they'll anyhow take it.
i. In a team, you are never absolved of responsibility. Whose job is it? It should have been someone else's, but if you want to achieve your objectives you have to go beyond your own job.
j. I want them to own up and add to the plan, but I am not willing to risk a complete deviation from my own objectives. If they go too far, I have to stop them sometimes.

CONFLICT MANAGEMENT

In teamwork, conflict is both inevitable and desirable. But at the same time, it is true that conflict has to be tactfully managed and circumvented. Researchers have identified four different kinds of tactics used to manage conflict. These are avoidance, accommodation, compromise and collaboration. The linguistic signs of accommodation, avoidance and collaborative tactics include soliciting all members' views and preferences, redefining compromise and recording problems, mentioning other problems to breakaway from the conflict situations, etc. Conflict management is also done through negotiation. Negotiation, in this context, predominantly insists on only two distinct processes among negotiators: Win-lose and win-win. The former focuses on the competitive aspect while the latter seeks to integrate parties and viewpoints. Linguistically, a win-lose orientation is manifest through explicit expressions of position, words that refer to debt, concession, winning and losing. A win-win process, however, is shown through elaboration of ideas, exploration of the other's ideas and re-evaluation or re-framing of one's own interests in the light of others. The following are some of the language patterns that are used generally in win-lose and win-win situations.

Examples of expressions that show either win-loss or win-win form:

Win-Lose Forms: These expressions show the positions we take up and competitive spirit.

We have always said we need our supply before we consider your demand

We'll be out-selling everyone soon

Win-Win Forms: These statements show attempts to balance needs, considering a number of other strategies, etc.

As long as I can stay within our budget, I can go ahead with this.

What if we justified the trip?

What feedback do we all need to have to ensure that this joint proposal passes?

ACTIVITIES

2. Recall any incident where you witnessed conflicting interests between two groups. Retrieve some of the expressions that express the following categories.

a. Avoiding
b. Accomodating
c. Collaborating
d. Compromising

3. Read the following statements and recognize the processes they reflect.
   a. I can understand things only from a "design" point of view.
   b. I want to clarify that all these things are central.
   c. We can talk and try to come up with some kind of agreement.
   d. Today was tremendous… there was good interface on how the issue affects all of us.
   e. You have made us think what we're doing and how. This has saved a lot of frustration and animosity.

> *When he took time to help the man up the mountain, lo, he scaled it himself.*
>
> *—Tibetan Proverb*

COMMUNICATION IN TEAMS

The previous discussion concentrated on the different kinds of teams, the nature of interaction between them and the linguistic identifiers to recognize the team type. Along with deciphering the talk, it is also important to understand the kinds of roles the team members play in a team. This is important because the role-type determines their talk in more ways than one. You must have noticed during any team talk that someone in the team initiates the discussion, some people add information and when the discussion gets serious, someone cracks a joke to relieve the tension. Some people with their general goodwill hold the team together, while some become the unappointed norm keepers.

Role-Taking in Teams: Task Roles, Building Roles, Maintenance and Negative Roles

> *If we were all determined to play the first violin we should never have an ensemble. Therefore, respect every musician in his proper place.*
>
> *—Robert Schumann*

It is important to note that the kinds of roles people take up will in turn, determine the kind of language they use and the kind of attitude they adopt. The roles people take up during a discussion can be divided into_**Task Roles, Building Roles, Maintenance and Negative Roles.** They can be summed up as the following:

Task Roles

Task roles include the communication functions necessary for a group to accomplish its task. Roles in which we perform different tasks in the team like the initiator, the information giver, the collaborator, the orienter and the evaluator are seen as task roles. These tasks can involve us in problem-solving, decision-making, exchange of information or conflict resolution.

Building and Maintenance Roles

These roles build and sustain the group's interpersonal relationships, helping everyone to feel more positive about the group's task and interact constructively and harmoniously. These roles include those that keep the team together like the harmonizer who compromises differences; tension reliever who induces humour; norm-keeper who ensures that norms are followed and the solidarity builder, who expresses and creates positive feelings reinforcing group cohesiveness.

Negative Roles

Intra-team behaviour may also be negative and self-centered in nature. These are the roles that are not constructive for the groups. They prioritize self-interest over group-interest. They are:

Blocker : a blocker constantly objects to ideas or suggestions, insisting that nothing will work. He may also repeatedly bring up the same topic or issue after the group has considered and rejected it.

Aggressor : this is the kind who insults and criticizes others and shows jealousy and ill will.

Storyteller : this is the one who tells irrelevant, often time-consuming stories and enjoys discussing personal experiences.

Recognition seeker : this is the kind that interjects comments that calls attention to his achievements and successes.

Dominator : this is the one who tries to monopolize group interaction.

Non-contributor : a non-contributor is reticent, uncommunicative and fails to respond to other's comments. He refuses to cope with conflict or take a stand on an issue.

Confessor : he is the kind who attempts to use the group as a therapeutic session and asks the group to listen to personal problems.

Special-interest pleader : he represents the interests of different groups and pleads on their behalf.

Activities

4. Imagine that you will have to make a team of all your acquaintances and friends for conducting a workshop. List down the names and depending on their predominant behavioural pattern, allot appropriate roles to each of them, thus ensuring optimum efficiency.

5. The following are some of the expressions that were used during an animated discussion. Identify them with the roles you think their speakers were enacting at that point.
   a. This is sheer non-sense. We cannot allow this. __________
   b. I think all of us have been speaking about... __________.
   c. I have a wonderful idea; why don't we... __________.
   d. I completely agree with you. __________.
   e. See, when I was the leader last time... __________.
   f. Let's not fight over it any longer. __________.
   g. How about giving the third party a chance? __________.
   h. I'm sure all of us will agree to this. __________.

GROUP DISCUSSIONS (GD)

> *Very often, conversations are better among three than between two, for the reason that then one of the trio is always, unconsciously, acting as umpire, interposing fairplay, seeing that the aggressiveness of one does no foul to the reticence of the other.*
>
> —*Christopher Morley*

A group discussion is an exchange of information, opinions, views, perspectives and ideas about a topic among members of a group. There is no particular number of participants that has to constitute the discussion group. But generally it has been found that in a group of ten or more some tend to avoid participation. But a group of five or less suffers from a lack of diversity of opinion and knowledge. Between five and nine can be therefore considered to be an appropriate number. As the discussion proceeds, one of the participants may emerge as a leader.

The expectation is that there will be smooth flow of information and interaction, and finally they can arrive at an agreed solution or strategy of action. A GD is an exercise and test of both your speaking and listening skills, in addition to being an experience of group dynamics in a face-to-face situation.

An effective GD is one where there is an equitable distribution of participation by all instead of being focused on individuals. Some participants may be very shy and inhibited, others may be very aggressive and dominating and a few others may be downright unruly, abrasive and disruptive. The success of a GD is ensured only if the members maintain certain decorum, discipline, harmony and balance.

GDs have become important in job selections as well as for admissions to professional courses. During the discussion, the candidates are generally judged for intellectual ability, creativity, and approach to problems, qualities of leadership, tolerance and group behaviour. In real life, group discussions contribute much to problem-solving.

To make a group discussion successful, it is important to pay attention to the following points:

- **Content**
  To make an impact in a group discussion, it is important to have a good knowledge of the topic given. It is important thus that you have fairly good general knowledge and awareness of the current situation. This will prevent the ideas from drying up fast and keeps the discussion alive and lively. If you are entirely unfamiliar with the topic given, wait for someone else to come up with important information and facts. Then quickly formulate your stance and come in with your perspective.

- **Communication**
  Along with the knowledge of content it is very important to know how you can communicate them effectively. It always helps to have a good grasp of vocabulary and fluency in speech. Using the right word at the right time gives clarity to the discussion and also highlights your role in generating ideas in the group. Remember not to exhaust your ideas at one go. Every time you contribute, make your talk brief and relevant. It is better to break into the discussion more number of times with a new idea each time rather than exhausting all your points the first time itself.

- **Thinking**
  Thinking is one of the important activities that have to go on during a group discussion. During a discussion you have to listen and understand the arguments of the other participants and at the same time decide what points you should raise and how. A good discussion always involves a lot of networking. And networking involves active thinking, building on one another's points, negotiating, persuading and collating views.

- **Group behaviour**
  Group behaviour is one of the qualities that is put to test during a group discussion. In fact group discussions are conducted to test initiative-taking abilities, leadership qualities and capacity to coordinate diverse viewpoints. Although expressing your views emphatically will be appreciated in a G.D, it is equally important that you draw the more reticent participants into the discussion and involve them in the decision-making process.

> *If everyone listens to each other's ideas, the truth will gradually and calmly emerge.*
> *—Jean Vanier*

STRUCTURING THE GD

Here are a few language tips that will help in structuring the group discussion:

Entering a discussion

Make comments on previous contributions and show your relation to them. You may have to change the trend of discussion by agreement, disagreement, amplification or by restricting the scope of the discussion.

Opening

We 're here today to discuss....
Let's decide how to proceed about the discussion.
Let's start off with no.1.
Can you please give your views on...?

Comments

What I think is...
I feel that...
The main point I wish to make is...
I agree up to a certain point but....
I must disagree with your opinion...
I would question that whether...
It seems to me that...
As far as I'm concerned...
I don't agree with the previous speaker...
Please, let me finish.
I think we are moving away from the main point.
If I may turn, now, to...
I want to comment briefly on...
I intend to make ... points about...
Now, to elaborate on the first point...
I strongly believe that...
With all due respect...
If we look at it in another light...
On the contrary...
I don't think any one could disagree with...
I can't help thinking...

People change and forget to tell each other.

—Lillian Hellman

Successful Group Discusion

A good and successful group discussion is one where the topic has been discussed threadbare. To ensure that, do the following:

- Analyze the topic word by word. Identify the frame of reference you would be using during the discussion.
- Look at the topic from the point of view of all the affected parties.
- Look at the topic from all the various angles and all possible perspectives.
- At the end of a discussion or when you know that the discussion time is up, it is necessary to give an appropriate conclusion. To do this, quickly try to recap the important points that have come up during the discussion, emphasize the points on which there were differences and where there was convergence of opinion and make the concluding remark.

Points to Remember

- Prepare well by reading and reflecting on the topic.
- Anticipate the points of others.
- Alertly listen and understand the points made by others.
- Break in and make your point without waiting to be called upon to do so, ensuring relevance to the context.
- Be loud enough to be heard by everyone.
- Make brief remarks often rather than giving long speeches and come in frequently into the discussion.
- Be open-minded and conciliatory rather than dogmatic.
- Avoid personal attacks and name-calling. Accept criticism with dignity and rebut it with solid points.
- Back your arguments with evidence and authority.
- Maintain eye contact with group members.

ACTIVITY

8. Given below are some topics for group discussion. See how many points you can build into each of them.

a. All the world is a stage.
b. Black is beautiful.
c. Deforestation is harmful.
d. Working mothers are more equipped to bring up children today.
e. Marriages are made in heaven.
f. The ceiling is high
g. Beauty lies in the beholder's eyes
h. The sky is blue

Try to bring around six people together and discuss these topics for around twenty minutes each.

INTERVIEWS

Viewpoint

The mention of interview creates nervousness in many people. There are not many who are undisturbed by the idea of facing an interview. In fact, in our day-to-day lives, we go through interview-like situations very regularly without being conscious of them. For instance, when sitting in a railway waiting room for your train to arrive, you may possibly start a conversation with the person in the adjacent chair. You may talk about your names, destinations, occupations, etc. When you visit your doctor, you are once again in an interview mode. The doctor asks for the details of your problems in order to diagnose and treat you. Many such situations happen all the while and we deal with them easily, quite free of tension. However, while facing a formal interview for admission or recruitment we are very conscious of the situation. Such interviews require a great deal of mental preparation.

Interview is a head-to-head interpersonal role situation in which the interviewer asks the interviewee specific questions with the purpose of assessing the interviewee's suitability for admission, recruitment, or promotion, or for an opinion. Hence an interview is a psycho-social instrument. It is an organized method of contact with a person to know his or her views. It is regarded as one of the important methods of data collection.

> *Interview is a systematic method by which one person enters more or less imaginatively into the inner life of another who is generally comparatively stronger than him.*
>
> *—P.V. Young*

Major Purposes of an Interview

1. **Collecting information Face-to-Face**
   An interview is a direct technique of accumulating data by the interviewer. Its advantage is that lots of information needed in social and scientific research or assessment can be compiled from the respondents, only on the basis of direct questioning.

2. **Hypothesis Formation**
   An interview is an investigative device to help identify variables and relations to suggest a hypothesis and to guide different phases of the research. This is relevant only to interviews for research.

3. **Value-Based Qualitative Facts**
   Social facts are essentially qualitative in nature. They are expressed in the form of ideas, feelings, views, faith, and beliefs. These social facts are both individual as well as collective. The interview method is an effective way to collect qualitative facts.

4. **Additional Information**
   By this method, we can accumulate from respondents, additional information that we often get through normal schedule or questionnaire. There are persons who are capable of providing additional information or suggestions and this can be done well through the interview method. This is possible as there is a direct dialogue with the respondents.

Significance of the Interview Method

1. **Getting Information about Feelings**: No other method of social research provides better information about a person's feelings, emotions and sentiments than an interview.

2. **Securing information from persons of varying levels**: The questionnaire method is useful only for the literate. Sometimes, questionnaires cannot elicit the required response from persons of different levels of intelligence. This disadvantage is surmounted by the interview method.

3. **Psychological study**: It is a psychological and scientific technique of observation which helps the interviewer to understand the person he is interviewing.

4. **Mutual motivation**: As a result of two or more persons coming in contact during an interview they motivate and enliven one another mentally. This helps the respondent to give the answers that are required of him. By observing the respondents closely, an interviewer can ascertain some aspects of the candidate's behaviour.

5. **Verification**: Information obtained through questionnaires and observation cannot be verified easily. However in an interview, it is possible to verify the information that has been gathered from sources like the CV or the application form.

TECHNIQUES OF INTERVIEWING

There are different types of interview techniques that are designed. Companies, government organizations, autonomous bodies, and private organizations shortlist candidates in a variety of ways. The different interview techniques are:

Series Interview

A company or a government organization selects its candidates through a series of interviews where the candidate has to face several people, individually, in succession rather than facing them all in a group. Each interviewer submits his/her report and the reports are compared to arrive at the final decision. The series interview is useful if there are a number of candidates to be interviewed for a job. It would be difficult for a panel of experts or senior managers to interview all of them in a limited time.

Filter Interview

In this, the applicants who do not have the minimum required qualifications are filtered out. If there are a large number of candidates who qualify, the minimum criteria are revised in order to recommend not more than a fixed number of candidates for further selection procedures. The filter interview may be done face-to-face, by phone, or some other method.

Set Interview

In such an interview, the interviewers ask all the candidates the same set of questions. It is a structured form of interview in which computers may be used to store, retrieve, and compare the data provided by the candidates. In a set interview, candidates do not talk to one another and do not disclose the questions.

Simulation Interview

The attempt in this is to simulate the conditions under which the job needs to be done. A situation is imaginarily created and the candidate is asked to demonstrate his/her skills and traits in dealing with the situation. The interviewer poses 'if' and 'when' questions to assess the candidate's reactions. Such interviews consist of just two or three situational questions. The answers need to be given carefully. It is better to express lack of knowledge about a situation than to pretend knowledge.

Spot-Calibre Interview

This kind of interview tests a candidate's mettle to find out how he would respond to difficult situations. In this, the chairperson after the initial formalities, indicates to a panelist to begin the interview. The candidate is subjected to stress by one or more of the following methods.

- being asked many questions at a time;
- being asked further questions without being given adequate time to respond;
- being interrogative in a dominating tone and voice, and
- being asked an irrelevant set of questions;
- being asked a provocative set of questions;

In this, it is better to stay calm and ask for time to answer the questions. The tone must be guarded and there must be no retort or retaliation or angry responses.

Panel/Board/Committee Interview

This is the most common type. An interview committee is set up with members from administration, finance, management and subject-specific experts. The chairperson of the panel conducts the interview with the help of the members and experts.

PREPARING FOR AN INTERVIEW

The following points must be borne in mind when presenting oneself for an interview.

Time Management

You must ensure that you are punctual. You must report for the interview at least half-an-hour before the given time. Punctuality creates a good impression and if you are on time you will be cool, calm and collected. If you find that you are going to be late for the interview, try to inform the concerned office so that the interview can be rescheduled. Be polite to everyone you meet both before and after the interview as this will also help you retain your balance.

Appearance

First impressions are immediately formed by our appearance. A smart and pleasant appearance enables you to gain a favourable impression from the board. It is essential to be well-dressed and well-groomed. The clothes you wear must look neither too casual nor uncomfortably formal. They must have a washed and ironed look. Footwear should feel comfortable and be polished.

Body Language

A candidate's body language tells us a lot about personality type. Stooping shoulders and bent forward back are signals that signify subservience and a lack of confidence. You should walk into the room and sit down with a straight back posture. Ask permission to sit by saying, "May I sit down?" Shake hands with a firm grip while maintaining eye contact and a smile. Your handshake is a basic act of friendliness. Therefore, a domineering handshake will irritate the interviewers who will subconsciously feel intimidated. Also, if your handshake is floppy, you could be considered a weakling.

A pleasing appearance with a little smile enhances your personality. A smile is like an antibiotic to the many of man's problems. A frowning, tired, or harsh expression can irritate the interviewer. You must therefore try to look cheerful and confident.

A good interviewee listens keenly, is alert and is able to draw the interviewer's attention as well. Attentiveness means being in tune with the other person's needs, wants and questions. It is being sensitive and treating each and every person in a special, distinctive way that recognizes his/her individuality. Good listening skills are vital if one wants an interaction to be effective because most people, in general, love to talk to those who listen to them keenly.

Points to Remember

- Do not sit in your chair in a very stiff or in an overly relaxed way.
- Do not get too near to the interviewer.
- Do not keep hands in your pocket.
- Do not cross arms.
- Do not get your hands or fingers over your mouth when you speak.
- Do not avoid a reasonable eye contact with the interviewer.

Communication Skill

Once you start speaking, it improves upon or negates the first impression made by your appearance and body language. Hence, effective speaking includes both content and delivery. Delivery refers to your tone, voice, choice of words and phrases. A candidate can make a better impression if the pitch of the voice is modulated, well based on syllable stress and context.

Enthusiasm

A candidate is given more quality attention from the board if he/she expresses more enthusiasm during the interview. Enthusiasm is reflected in the energetic way you express your ideas. You should never be laid-back or playful, but maintain all along a cheerful disposition and a pleasant appearance.

Brevity

Effective communication does not mean speaking in a garrulous and flowery language. Authentic communication means speaking briefly with clarity and in an unambiguous way. A talkative person is often taken lightly.

Listen Carefully

Very often people do not exercise patience to listen to others as they are anxious to speak or express themselves. Listening is the mother of all communication as it pleases the speaker as well as orients you positively. You can more so understand the question properly and give appropriate response. Try not to speak before the question is completed.

Be Honest

You should never attempt to deceive the interviewers by telling lies. If you do not know the answer, it is correct to acknowledge it. You gain respect for your integrity and honesty. You can tactfully steer the interview to areas that are familiar to you.

Adapt Yourself

It is always useful to keep the interest of the interviewer intact. An element of enterprise is always helpful in facing an interview. You must have an idea about the interests of the interviewer and talk about it. If the interest of the interviewer seems to decrease, you can trigger his interest by making appropriate change in your tone, by lowering or raising your voice, or by speaking faster or slower.

Maintain Proportion and Poise

You should answer informatively but not consume much time. The interviewer may ask general question such as, "I would like to talk a little about your home background". Do not give monosyllabic answers or ramble. Yet, do not talk so much that you may weaken your candidature.

Demonstrate Leadership Qualities

You ought to exhibit initiative, willingness, and resourcefulness. You ought to demonstrate team spirit, cooperation, organizational skill, strength of character, and more importantly, a strong decision-making ability. Diffidence, selfishness, and a withdrawn nature will be counterproductive.

Points to Remember

- Walk in smartly and cheerfully.
- Shake hands firmly without crushing.
- Maintain a reasonable amount of eye contact with the members.
- Give him your full attentive listening and observation.
- Modulate your voice such that your reply is audible to every member of the interview board.
- Neither raise your voice rudely nor speak in a low tone.
- Do not be in a hurry to answer. Speak distinctly in normal accent and pause to make points effectively.
- Do not move your limbs aimlessly. Restrain their movements to the minimum.

KINDS OF QUESTIONS EXPECTED AT INTERVIEWS

Alec Rodger came out with a discussion on interviews and said that interview questions fall under seven headings. These are (1) physique, (2) attainments, (3) general intelligence, (4) aptitudes, (5) interests, (6) disposition, and (7) circumstances.

Physique consists of health, appearance, manners and other related aspects of personality. Attainments include educational achievements and experience. General intelligence is seen as a broad area of common sense. Aptitudes comprise mechanical, verbal, musical, or artistic skills. Interests comprise outlook and hobbies. Disposition includes personality and related aspects like acceptability, whether you are self-motivated, whether you are dominant or submissive, extrovert or introvert, etc. Circumstances give a pointer to the interviewer to put your achievements in a perspective.

Nonetheless, the expected questions fall within four groups:

Leading Questions

These generally have a response built into them. Normally, the chairperson begins the interview with these leading questions. An interviewer is usually friendly and cooperative; he would like the candidate to feel at ease. Such questions help a candidate feel comfortable and confident to do well.

Open-Ended Questions

These questions usually come after the leading questions. The purpose of these questions is to help a candidate to talk, explain, and illustrate something that he/she knows or has done before.

Probing Questions

These questions measure your depth of understanding. These questions are intended to test how you would react to a situation or how you would organize follow-up questions.

Close-Ended Questions

These questions try to elicit information on specific items and to test your knowledge of facts and figures.

THE INTERVIEW PROCESS

Step 1: Initiation of the Interview

The interviewer's task is not to bamboozle the candidate but to get the best out of him. Normally, therefore, the interview begins with encouraging, lively questions. There are many methods of initiating an interview; some of them are discussed below.

Initiation based on candidate's background : In many cases the chairperson tries to begin the interview with questions relating to the educational or the family background of the candidate. He may enquire about the place the candidate belongs to, its important places, features or persons. The purpose is to break the ice, make the candidate feel at ease and to make the process of interview interesting for the candidate.

Initiation based on the candidate's interests and hobbies : The chairperson could put a question relating to the candidate's field of interest. Candidates must be honest in mentioning their hobbies and interests. Wrong information could lead to a question, the answer to which may not be known to the candidate.

Initiation based on general knowledge : Sometimes the member may initiate the interview with questions that test the general awareness of the candidate. This is a tough beginning. Questions may be asked on subjects from the evolution theory to the www, from nuclear proliferation to the current visit of the PM, from global climate to the problem of living styles etc. Thus, candidates must work hard and possess wide knowledge on a variety of subjects. They should read newspapers and magazines regularly, listen to news, and discuss current affairs with friends and relatives regularly.

Initiation based on academic topics : This method is actually convenient for the candidates because the questions asked are relating to subjects of specific interest to the candidate. He/she should, therefore, be thorough with the basics, fundamental concepts, and the latest information regarding the discipline concerned. He/she must develop the confidence to explain or clarify any related question.

Initiation based on tricky questions : Sometimes, the interview can begin with perplexing questions. Such questions are meant to measure the ability of the candidate in an awkward situation. Candidates ought to remain calm under such circumstances.

Step II: Investigating the Tenor of Behaviour

Since an interview is an appraisal of the total personality of the candidate, it is essential that the interviewer investigates the implications of the behavioural tenor of the candidate right from beginning to the end of the interview. The dialogues between the interview members and the candidate also reveal many aspects of the candidate's personality.

Step III: Assessing the Candidate's Knowledge and Understanding

The board evaluates the candidate's general knowledge, study of specific subjects, understanding relating to current affairs, interest in and critical awareness of all things happening around. The candidate's ability to apply this knowledge in a given situation or for a social problem is also tested. His ability to organize ideas and information into a coherent concept or approach is also tested. Candidates must make a rigorous study of their subjects and have an up-to-date knowledge of current affairs. For this, they should read editorials and important articles from magazines and newspapers, and go through the analyses ov various topics, either in newspapers or on television.

Step IV: Assessment of Interpersonal and Social Qualities

Generally, the questions asked at different stages of an interview themselves reveal the social personality of the candidate like sense of responsibility, co-operation , adaptability, integrity, group work and persuasiveness. However, the interview board may also ask specific questions which will reveal the above qualities of the candidate.

Step V: All-in-All Assessment

In this last stage of the interview, the board generates a final impression of the candidate. Therefore, some questions may get repeated to judge whether the candidate is consistent and firm in his attitude.

Norms for Assessment of a Candidate

An interview is basically an assessment of the overall personality of a candidate. Personality does not just mean mere outward appearance or intelligence or the ability to get going in a situation; it is a blend of various qualities of mind, body, and spirit. Personality can be categorized under four heads: 1. disposition, 2. knowledge, 3. communication skill, and 4. leadership traits. These four heads can be further classified as given below:

1. **Disposition**
   a. Appearance
   b. Social manners
   c. Dynamism
   d. Mental power
   e. Overall impression
2. **Knowledge**
   a. Range of knowledge
   b. Depth of knowledge
   c. Application of knowledge to real situation
   d. Coherence of thought
   e. Overall impression
3. **Communication skill**
   a. Language
   b. Voice, tone, rhythm
   c. Clarity and logic
   d. Convincing power
   e. Overall impression
4. **Leadership traits**
   a. Initiative
   b. Organizational skill
   c. Deciding power
   d. Character
   e. Overall impression

Each basic head can be allotted marks. Negative marking can also be done.

SUMMARY

- Teams work together largely through conversations.
- All teams have in-built in their structure, both independence and interdependence.
- Team members have to trust one another, while remaining vigilant at the same time.
- Acknowledgement of mutual interest, proposals for joint action and soliciting others' views and preferences for taking independent decisions are indicative of the interdependence or independence of the team members.

- Certain language forms like calling one another by their nicknames, using informal pattern of speech, sharing views and showing concern for one another's interests and wants, express closeness among team members.
- Task accomplishing, building a group's interpersonal relationships and maintaining balance in a team are some of the positive roles of individuals in a team.
- A blocker, an aggressor, a storyteller, a recognition seeker, a dominator, a non-contributor, a confessor and a special-interest pleader play negative and self-centered roles within teams.
- A group discussion is an exchange of information, opinions, views, perspectives and ideas about a topic among members of a group and is a test of both speaking, listening skills and group dynamics.
- It is important to network one's ideas, build up the argument and give the discussion a definite direction.
- The appropriate number of participants for a GD is between five and nine and is effective when there is an equitable distribution of participation by all instead of being focused on individuals.
- The success of a GD is ensured only if the members maintain a certain decorum, discipline, harmony and balance. Good listening forms an important part of group discussion.
- GDs have become very important in job selections as well as for admissions to professional courses.
- To make a group discussion successful, it is important to pay attention to the content, communication, thinking, group behaviour and ensure that the topic has been discussed threadbare.
- Interview is an interpersonal role situation with the purpose of assessing the interviewee's suitability for admission, recruitment, or promotion, or for an opinion
- Interview is a psycho-social instrument.
- Preparation for an interview involves improving aspects relating to one's time management, appearance, body language, communication skills, enthusiasm, brevity, listening ability, honesty, adapting power, tendency to maintain proportion and poise and demonstrating leadership qualities.
- Interview questions fall under seven headings—physique, attainments, general intelligence, aptitudes, interests, disposition, and circumstances.

  Interview questions may also be classified under four groups — leading questions,open-ended questions, probing questions and close-ended questions.

REVIEW QUESTIONS

1. State the inherent and inbuilt contradictions that are present in every team. Explain each of them with examples.
2. What are the different parameters we can use to judge a team? Give a short account of each.

3. Give a detailed account of power differentiation. Show the variables that signify power and explain with illustrations the speech types that co-relate with each of them.
4. What are the different ways in which teams manage conflict? Give illustrations to explain each method and show the language forms that go with them.
5. How do interdependence and independence of team members get conveyed?
6. How does closeness among team members manifest itself?
7. What are the different team types?
8. Discuss the term 'collaborative team'.
9. What are some of the positive roles of individuals in a team?
10. Who are the typical negative role players in a team? Discuss.
11. What is a group discussion and what is its ideal number of participants?
12. What is the importance of turn taking during a group discussion?
13. What is an interview?
14. How should a candidate prepare for an interview?
15. What are the different interview techniques?
16. How many types of interview question have been identified in the chapter?

CHAPTER 10

Telephone Skills, Meetings and Minutes

Chapter Objectives

This chapter discusses the important aspects relating to telephoning, conducting of meetings and minutes writing. It focuses on the do's and the dont's of these processes. The purpose of the chatper is to familiarise students with these concepts and enable them to use it when necessary.

TELEPHONIC COMMUNICATION

Telephones are a common means of oral communication. It is not always possible to have face-to-face contact with the personnel within and outside one's organization. But it is easy to contact one another on the telephone. Planning telephone calls is a simple, yet important activity, which does not take much time. Planning calls saves speaking time and money. We can always prepare an item list and have the relevant papers at hand, before making a call.

Some points with regard to telephoning

1. Positive (Do's)

- Calls are to be answered promptly.
- Pen and message pad have to be kept at hand.
- Statements and replies are to be given with the appropriate information. E.g. Extension 227, Ravindran speaking.
- Responses are to be spoken into the mouthpiece pleasantly and distinctly.
- Voices of people we know are to be recognized and their names used or the caller's name has to be obtained politely.

- People who may have to be put on hold on the phone while waiting for information have to be assured periodically that their concern would be addressed at the earliest.
- Callers have to be given the chance to ring back instead of being placed on hold for a prolonged period..
- Messages have to be taken and written out correctly so that the recipient will be able to read them easily. The messages have to be kept at a place where they will be seen.
- When callers become annoyed or impatient, the receiver should exercise self-control.
- Knowing enough about one's organization helps the employee to re-route calls that come to him by error.
- Instructions are to be left, when an employee is expecting a call during his absence.
- During a call, distractions are to be avoided.
- Cultivate a lively telephone voice. Find out how you sound from your friends or relatives.
- A lot of care has to be taken while speaking to foreigners.
- Be very clear and distinct while giving names and numbers.
- If the person the caller wants is not available, offer a return call.
- Before calling anybody, ask yourself if it is the right time to call and why you want to call.
- Before making a call, jot down the points to be made and keep all the necessary documents at hand.
- Have a clear idea of the message for the person you are calling.
- When someone answers your call, greet him or her and identify yourself.
- Apologize, in case you get through to a wrong number.
- When you get through to the right person, check if it's the right time to talk to him or her.
- Re-dial if the line gets disconnected. When the call is complete, thank the person you have spoken to.
- After making a call, jot down the key points of the conversation for future use and also to avoid forgetting them.

2. Negative (Don'ts)

- The instrument should not be misused.
- Slang expressions should be avoided.
- A conversation should not be carried on with someone else in the room while attending a call.
- A person who is wanted on the telephone should not be called across the room unless, the mouthpiece is covered.
- Do not ramble during a call. Be considerate to the other person, as he or she may be in a hurry.
- Before answering a call, don't let the phone ring more than two or three times.
- Do not allow the phone to interrupt a meeting or anything equally important.

On the phone: A few patterns

1. Ram: Dials/connects/
(Phone rings)
Operator: Hello, NTPC.
Ram: Hello / Good Morning / Good Afternoon /
I am Ram here. May I speak to Zaheer?
Could you put me on to Zaheer?
Could I speak to Zaheer?
Operator: Yes sir, please be on the line.

2. Ram: Hello, Ram here. May I speak to John, please?
John: Speaking.
Ram: Good Morning, Mr. John.

3. Ram: Hello, This is Ram speaking. Could I speak to John please?
Operator: Just a moment please. I'll put you through.
John: Hello, John here.

4. Kumar: Hello! Could I speak to Ram please?
Operator: Sorry, there is no Ram here. May be it's a wrong number.
Kumar: Oh, I'm sorry.
Operator: It's all right.

MEETINGS

> Some people must learn that a meeting of minds and a bumping of heads are quite different things.
>
> —*Dell Pencil Puzzle and Word Games*

A meeting may be defined as a formal gathering of a group of people at a pre-arranged time and place. Though generally formal in nature, a casual and informal meeting with no prior notice is also possible. Meetings are generally held for the following purposes:

- Information gathering
- Information giving
- Persuading
- Problem solving
- Decision taking
- Settling disputes
- Formulating policies

Guidelines for Effective Meetings

- Circular or information has to be sent to all participants in advance about the time, place and probable length of the meeting.
- The **agenda** of the meeting should also be circulated. The topics to be discussed may be placed in order and the last item should be listed as AOB (any other business with the permission of the chair).
- As far as possible, the purpose of the meeting should be achieved through negotiation, consensus or majority vote. If the participants are unable to agree, it must be recorded.
- Notes of the meeting should be made for preparing the minutes.

Conducting a Meeting

If the responsibility of conducting a meeting falls on you, you will need all the skills at your command to maintain control and create the right atmosphere for discussion. You should start the meeting with a summary of the objectives of the meeting and define the terms and scope of discussion. This will help you to keep the discussion relevant and on track. Very often, participants tend to cross talk or indulge in private discussion in groups. Differences of opinion may emerge and tempers do get lost. It is necessary for you to be calm and restrain the over-enthusiastic members. This can be done by ensuring that everyone gets a chance to speak, particularly the silent ones. The meeting should be concluded with a final summary that reiterates the conclusions of the meeting and provides information about the date of next meeting. Make sure that someone takes down notes to prepare the minutes. Firmness but openness with sensitivity to individual speakers and listeners are the hallmarks of an effective chairperson.

Participating in a Meeting

Be sure of your facts and information. Be clear in your mind about what you want to say and say it with clarity and confidence. It is true that due to power equations, it may not be possible to be forthright but you may express yourself by giving your reasons for your point of view.

If you have to interrupt other participants, do so politely. The same applies when you are contradicting or criticizing others.

The Language of Meetings

The following are some of the phrases generally used during meetings:

Chairing a meeting

- Good morning/afternoon/evening
- I welcome you all to this meeting
- The copy of the agenda is in your hands.
- We shall begin with…
- ____, would you like to begin with…
- Your attention is drawn to…
- I particularly wish to draw your attention to…
- Are we all agreed that…?
- We seem to be losing sight of the main point…
- We seem to be getting sidetracked
- We can talk about it later when we have got all the information.
- I should like you to vote in this issue.
- If no one has any objection, I propose that we leave this matter in abeyance till we can discuss this further.
- I'm afraid that we'll have to end here.
- To summarize, we decided that…
- Thank you all for coming.
- I would like to thank you all for your cooperation.
- That concludes our business for today and I declare the meeting closed.

Taking part in a meeting

- Excuse me, Mr Chairman, may I say something.
- May I comment on what…has just been said?
- Could I say something here, please?
- Could I make a point about…?
- May I add to what he has just said?
- It sounds like a good idea but….
- That doesn't sound to me like a very good idea…
- Let's not forget that…
- Are you sure that…
- I should like to remind the gathering that …
- Perhaps I haven't made myself clear…
- Under no circumstance can…
- I agree to…

While refusing ice cream after lunch, a professor told the students that he had a health problem. After a pause, he said it all, "Some people have dental problems, and some others have transcendental problems."

MINUTES WRITING

A report covering the proceedings of a meeting of a business organization is known as 'minutes of the meeting'. The minutes of any meeting – whether it is a company's board meeting or a managing committee/executive committee meeting of an association – are written by the secretary of the organization. As the minutes represent a legal record of the proceedings of meeting, the decisions taken are recorded in the form of resolutions. Personnel references are avoided as far as possible. Passive voice sentences are generally preferred.

E.g. Instead of saying: "Mr Bhujang's proposal that a new storeroom should be built for the corporation was passed", in the minute book it is recorded as "Resolved that a new storeroom be built for the benefit of the corporation".

If the minutes are minutes of narration, personal references are included but not when the matter is of a defamatory nature. In the minutes of decisions, no names are mentioned.

Minutes have to be correct. Normally resolutions are numbered in the serial order and the serial numbers are carried forward from one routine board meeting to another. Formal changes of the office-bearers have also to be recorded. Misrepresentation of facts regarding the proceedings of a meeting is an offence and hence the normal custom is to read out the minutes of a meeting at the beginning of the next meeting and get them approved by the members before the chairman signs them.

Pattern of Minutes

LETTER- HEAD OF THE COMPANY/NAME OF THE COMPANY

Day and date of the meeting:
Venue and the kind of the meeting:

Members present:

Mr ____________________, Chairman
Mrs ____________________, President
Mr ____________________, Director
Ms ____________________, Treasurer

In Attendance:

Mr ____________________, Secretary
Mr ____________________, Solicitor

On the agenda:

a. ____________________
b. ____________________
c. ____________________
d. ____________________

No.	Title	Details
1.	Minutes of the previous meeting	____________________
2.	Transfers	____________________
3.	X	____________________
4.	Y	____________________
5.	Vote of thanks	The meeting ended with a vote of thanks to the Chair.

Date: **Chairman**

The following are the important features of a well-written minutes:

1. The minutes are generally written on the letterhead of the company.
2. At the beginning of the minutes, there are details such as the day and date on which and the time at which the meeting was held, the venue of the meting and the kind of meeting it was supposed to be.
3. This is followed by a list of those who were present at the meting.
4. The first resolution is normally the confirmation of the minutes of the previous meeting and the last resolution is normally the formal vote of thanks to the Chairman.
5. If any member/Director is absent and has written for permission, this resolution comes second. But if a condolence resolution is to be passed, it comes after the approval of the minutes of the previous meeting.
6. When an appointment is recorded (individual/group/sub-committee) the sentence says that the person or the committee "is hereby appointed". Similar wording is used when someone is authorized to do something in the name of the company.
7. At the end of the minutes, there should be provision for the date and the Chairman's signature. This data is the date on which the minutes are confirmed and signed.

Activity

1. Draft the minutes of the monthly executive committee meeting of your cooperative society at which, among other things, the following items were on the agenda:

a. To consider the proposal for increase in the loan amount.
b. To consider the need for acquiring a permanent structure for the society.

Minutes of Routine Board Meetings of Companies

Minutes of the routine meetings of Boards of Directors of Companies are guided by the agenda that is given. Some routine activities besides the agenda can also be recorded in the minutes, if these have been undertaken during the meeting. Prominent among these are considerations of applications for transfer of share certificates etc. The major items on the agenda are normally given in advance. If the agenda contains proposals that cannot be subject to spot-decisions, committees are appointed for further investigation or study of the problem. When committees are appointed, the convenor and the members are mentioned by names, and the date by which they have to report to the Board is specified.

Sample 1: Minutes of a Routine Board Meeting

Minutes of the monthly board meeting of Bibliophile Publications Ltd. at which, among other things, the following items were on the agenda:

a. To consider the proposal for extension of the administrative office building of the company at Narayanguda.
b. To consider the request for building a welfare center for the employees of the company.
c. To consider the suggestion of appointing a Liaison Officer for the company.

BIBLIOPHILE PUBLICATIONS LTD.

Minutes of the Routine Board Meeting of the Directors of Bibliophile Publications Ltd, held on February 17, 2003, at 4-00 p.m. at the registered office of the company in Hyderabad.

Present : Mr Madhusudan
Mr Janakiram Rao
Mr Adinarayan Rao
Mr Muniraj

In Attendance: Mr C. Thrinath (Secretary)

No.	Title	Details
1.	Minutes of the Previous Meeting	The minutes of the previous board meeting were read, approved and confirmed.
2.	Share Transfers	Resolved that the request for transferring share certificates (no.9076 to 9087) be granted
3.	Extension of Administrative Wing	Resolved that the proposal for extending the administrative office of the company by taking over the recently vacated office at Balaji Towers be accepted. The Secretary was authorized to negotiate for this in consultation with the solicitors of the company.

4.	Construction of Welfare Center	Resolved that a sub-committee is for employees: hereby appointed to find out the feasibility and the cost of construction of a welfare center for the employees of the company. Further resolved that Mr Satish be designated the convener of the committee and Mr Kumar and Mr Gaurav be members of the committee and that the committee should submit its report by March 15, 2003.
5.	Liaison Officer	Resolved that a Liaison Officer be appointed at the head office of the company and that the Secretary is hereby authorized to advertise for the post.
6.	Next Meeting	Resolved that the next meeting of the board will be held at 6.00 p.m. on March 30, 2003 at the same venue.
7.	Vote of Thanks	The meeting ended with a vote of thanks to the Chair.

Date: **Chairman**

Sample 2: Minutes of the Meeting of Board of Directors (meeting held before the Annual General Meeting).

The board of directors holds a routine meeting before the annual general meeting to prepare for the annual general meeting. This meeting is known as the meeting of the board prior to the annual general meeting. Among the most important items of business transacted at this meeting, will be those recommending the dividend deciding on the date of the annual general meeting, deciding on the dates on which the register of the company will be closed, etc.

(Minutes of the Board Meeting (before the Annual General Meeting)

PRIME ATTIRES LTD

Minutes of the meeting of the Board of Directors of Prime Attires Ltd. held on August 27, 2003, at 5-00 p.m. at the registered office of the company at New Delhi.

Present:

Mr Kishore
Mr Jaisimha
Mr Vijayachander
Mr Osman Ali
Mr Aditya, Secretary.

No.	Title	Details
1.	Minutes of the Previous Meeting	The Secretary read out the minutes of the previous meeting and the same were confirmed and signed.
2.	Leave of Absence	Resolved that the leave of absence requested for Director Mr Reddy be granted.
3.	Annual Accounts and Balance	Resolved that the annual accounts and the balance sheet of the company as audited and prescribed at the meeting be approved.
4.	Books of Transfer	Resolved that the share transfer books of the company be closed from August 25 to September 4, both days inclusive, for year end work. Further resolved that the Secretary should arrange to announce this officially to the members.
5.	Annual General Meeting	Resolved that the annual general meeting of the company be held on September 29, 2003 at 4-00 p.m. at the registered office of the company. Further resolved that the Secretary should inform the members about this and arrange to send copies of the accounts and the Director's report along with the notice.
6.	Dividend	Resolved that after providing for general reserve fund an amount of Rs.__________ lakhs from the profits for the current year, a dividend fund of 10% be recommended to the members as the dividend on equity shares of the company.
7.	Director's Report	Resolved that the draft copy of the Director's report submitted at the meeting be adopted as it is.
8.	Vote of Thanks	The meeting ended with a vote of thanks to the Chair.

Date: **Chairman**

SUMMARY

- Telephones are a common means of oral communication
- Planning telephone calls is a simple, yet important activity, which does not take much time.
- Planning calls saves speaking time and money.
- We can always prepare an item list and have the relevant papers at hand, before making a call.
- A meeting may be defined as a formal gathering of a group of people at a pre-arranged time and place.
- Though generally formal in nature, a casual and informal meeting with no prior notice is also possible.
- Meetings are generally held for various official and semi-official purposes.
- Certain guidelines for conducting effective meetings including preparing an agenda, should be prepared.
- In a meeting, an effective chairperson has to be firm but open and sensitive to individual speakers and listeners.
- A report covering the proceedings of a meeting of a business organization is known as 'minutes of the meeting'
- Minutes writing is an important aspect of meetings in modern business

REVIEW QUESTIONS

1. What are the essentials of an effective telephone call?
2. What are the benefits of a telephone?
3. What are some of the do's and don'ts of telephone conversation?
4. What is a meeting and an agenda?
5. What are the purposes of conducting meetings?
6. What is meant by 'minutes of a meeting'?

CHAPTER 11

Business Letters, Technical Writing, E-mail Writing

Chapter Objectives

This chapter discusses most of the significant aspects relating to business letters, technical writing, and electronic mail (e-mail) messages which have become popular alongside other skills and important features of today's world.

BUSINESS LETTERS

Business letters are letters written in the context of business transactions between individual business personnel or between business organizations. It is essential to familiarize yourself with business letter formats and styles in different business situations and to gain some practice in writing a letter in a given business context. The other formats covering letters of inquiry, request, replies, orders and business reports. Different forms of layout like the indented form, hanging indention, block form, modified block and semi-block are also explained.

A visitor to a certain college paused to admire the new Hemingway Hall that had been built on campus.

"It's a pleasure to see a building named for Ernest Hemingway," he said.

"Actually," said his guide, "it's named for Joshua Hemingway. No relation."

The visitor was astonished. "Was Joshua Hemingway a writer, also?"

"Yes, indeed," said his guide. "He wrote a check."

Business Letters: General Layout

1 115, Right Street
Mayanagar
Hyderabad – 500 029
12 Aug, 2002

2 Mr R Shermani
Manager
Bank of Baroda
Hyderabad.

3. Dear Sir,

4 ____

5 I hope to receive an early reply

I am,

Yours faithfully,

Signature

(Name In Bold Letters)

The general layout of a well-written letter has the following important features:

1. Heading

The address, the date, <u>no name</u>

a. 115 Right Street
Mayanagar
Hyderabad-20
August 12, 2002

Presently the open-punctuation (no commas, full stops etc.) system is widely followed.
Preferred American style

August 12, (Date without th, nd, rd followed by a comma) (th, nd, rd are pronounced but not written)

OR

8-12-2002

OR

8/12/'02

b. August 12th, 2002 Preferred British style

OR

12/8/02
(th, nd, rd are usually written)

c. 12 August, 2002 Preferred international style (the date is placed first and this style is followed all over the world)

d. All the styles are acceptable.

e. A full stop is used after St. and Rd. – short forms for 'street' and 'road'.

2. Addressee's name above 'Dear'.
If you are addressing a person by name give the official designation also; if one is applying for a job then only the designation is given and not the name. In business letters, after the name of the person 'Esq' may also be added: R. Shermani, Esq. 'Esq' is an abbreviation for 'Esquire'. Never write both Mr and Esq:
E.g. Mr R. Shermani Or R. Shermani, Esq.

a. "Esquire' (British English) is used as a title of politeness usually written in its shortened form after the full name of a person
e.g. Kerry Packer, Esq.
Titles and decorations awarded by the State are placed before university degree.
e.g. Mr C. M. Murali Mohan, M.P., Ph.D.

b. 'Messrs' (Abbreviation for 'Messieurs' which means Gentleman)
is not used when addressing a company whose name is not personal.
e.g. The General Trading Co. Ltd.
but Messrs Shiv & Co. Ltd.

c. In the Indian context, some people use Smt/Sri/Kum in their correspondence.

3. The Salutation

Dear Sir,	My Dear Sir
Dear Gentlemen,	My Dear Madam,
Dear Madam,	Dear Mesdames,

1. If the name is not known, 'Dear Sir' is used.
2. If the letter is addressed to a woman, 'Dear Madam' is used.
3. If the letter is addressed to an organization or firm, 'Dear Sirs', or 'Dear Gentlemen' is used.
4. If it is addressed to an all-women organization, use 'Dear Mesdames'.
5. In a general circular, use 'Dear Sir or Madam'.
6. In formal official letters, one may use 'Sir'.
7. 'My dear Sir' is rather old-fashioned.
8. Official designations after the title may be used in the salutation in letters addressed to persons in their official capacity;

   Dear Mr President,
   Dear Mr Governor,
   Dear Madam Secretary,

9. Titles other than Mr, Mrs, Miss, Dr are spelt in full
   Dear Professor Rao,
   Dear Ambassador Kumar.

4. The Body

Common openings in business letters:

1. Thank you for your letter of Aug. 12…
2. Received your letter of Aug. 12…
3. I should like to order …
4. I should like to inquire whether …
5. I wish to …
6. I wish to apply for the position of … advertised in the …
7. I wish to complain about …
8. I have to complain about…
9. Would you please …
10. I wonder if …

Formerly, most business letters had a formal style. However, modern business letters are less formal. Avoid the chatty, casual element of the personal letter, but do not use the stiff, impersonal, distant tone that was once adopted for all official and business letters.

The body of the letter may be written in paragraphs. The first line of each paragraph is indented, unless the writer follows open punctuation.

Business jargon, once so popular, is now giving way to plain English;

Business Jargon	Plain English
Inst., or instant	of this month
Ult., or ultimo	of last month
Prox., or proximo	of next month
We are in receipt of your esteemed communication	We have received your letter
We hereby beg to inform you	We wish to inform you
'Re'	With reference to

5. Common Endings or Subscription

e.g. Yours faithfully	(British style in formal business letters)
Yours sincerely	(British style in letters between people who know each other by name)
Yours truly	(British style in letters where there is no personal relationship between the writer and the addressee)
Yours cordially	(British style in a formal letter where there is a desire to sound respectful)

'Y' in Yours must be a capital letter.

- In most cases, American usage prefers to put the adverb before 'Yours'; Sincerely yours, Cordially yours. ('Yours'; Sincerely yours, Cordially yours. ('yours' in this case does not begin with a capital letter.)
- 'Yours' does not have an apostrophe. (Your's is wrong).
- Sometimes 'I am', 'I remain', 'We remain' are used.

e.g. I am,

Yours faithfully,

Signature

- Avoid ending a letter with a present participle construction
  e.g. Hoping to receive an early reply - (wrong) (unacceptable)

  I hope to receive an early reply (acceptable)

I am (Open punctuation)

Yours faithfully

BUSINESS LETTER: SAMPLES

1. A Letter of Inquiry

123, Chickal Road,
Cuttack 795 001
12 March 2002

Vijaya Furniture Company,
Cuttack 795 361.

Dear Sirs,

From your advertisement in the *Indian Express* of March 4, I understand that you supply wooden and steel furniture in various latest designs.

I would be grateful if you could send me a brochure of the furniture you supply. I am particularly interested in buying furniture for my drawing room

Yours truly

Sd/-
(A.S.KELUCHARAN)

Writing is a kind of therapy. Sometimes I wonder how all those who do not write, compose, or paint can manage to escape the madness...which is inherent in the human condition.

—Graham Greene

2. Reply letter

VIJAYA FURNITURE COMPANY
CUTTACK 795 361.

Telephone: 8621
Telegram: VIJAFUR

20 March 2002

Mr A. S. Kelucharan
123 Chikal Road,
Cuttack 795 001.

Dear Sir,
Thank you for your letter of March 12, asking for a brochure of the items of furniture we can supply.

We are happy to send you herewith details of the wooden and steel furniture available with us. Different designs of furniture for the drawing room are shown on page 14 of the brochure. You have a choice of teakwood and rosewood for these items of furniture.

Wooden furniture can be supplied in a week. The delivery for steel furniture will take up to three weeks.

If there is any other information you wish to have, please write to us. We shall be happy to let you know.

Yours faithfully,
For VIJAYA FURNITURE COMPANY

(XXX)
Sales Manager

3. Placing an Order

VIJAYA FURNITURE COMPANY
CUTTACK 795 361

Telephone: 8621
Telegram: VIJFUR

Our Ref:

Your Ref:

2nd April 2002

Messrs. Azeem Brothers,
Timber Merchants,
Hyderabad – 500 021.

Attention: Mr Abdul Azeem

Dear Sir,

Sub: Our Quotation for teakwood planks

Your letter containing details of your teakwood stock was received today.

We think the teak wood sizes that you can supply will not meet the needs of our customers. Could you, therefore, send us details of planks in plywood you can supply within a month?

A full statement of the sizes and number of planks we need is enclosed. We would be happy to hear from you soon.

Yours faithfully,
For VIJAYA FURNITURE COMPANY,

Sd/-

(P RAMA RAO)
Managing Partner

Encl:
Copy to: Incharge, Carpentary Section

This letter has some extra features. Top left, we find two reference numbers. When there is a long correspondence on a subject and the mail has to be filed for future reference, reference numbers are given. With the help of the reference numbers, the party concerned will be able to trace the previous correspondence on the subject and gather the information required for taking decisions. After the inside address we see another feature, namely, the attention line. Here the name of the person incharge of the subject is mentioned. In a large firm with several sections, the attention line will ensure that the letter reaches the right person in time. Next to the signature at the close of the letter, the number of enclosures with the letter is mentioned at the left hand margin. We can mention either the number of papers enclosed, or give a list of the enclosures.

e.g. Encl: 1. Brochure of items of furniture in stock

If a copy of the letter has been sent to anyone else, that is shown below the list of enclosures.

e.g. Copy: Incharge Head, Carpentry Section.

CHARACTERISTICS OF A GOOD BUSINESS LETTER

- A business letter has to be courteous and considerate.
- It has to be precise and clear.
- It has to be complete.
- It has to be brief.

Read the following model letters. Letter A and Letter B are written by two different secretaries of a company. In both the letters, the purpose of the writers is the same, that is, company sends donation to a social organization.

LETTER: A

RADHA IRON AND STEEL INDUSTRY
BB 276, WALKER TOWN, SECUNDERABAD, A.P. 500 032

Tel. 27833056 Ref:

Date: 8/3/03

The Secretary,
R. K. Mission Seva Samstha,
Domalguda,
Hyderabad – 009.

Sir,

We came to know from your notification inst., about your need for funds for a cancer ward in your hospital. As you do social service, you need help from business people like us. We are sending a cheque for Rs.10,000/- in this connection.

Yours faithfully

Sd/- XXX

(Rakesh Bhimani)
Secretary
Radha Iron and Steel Industry

LETTER: B

RADHA IRON AND STEEL INDUSTRY
BB 276, WALKER TOWN, SECUNDERABAD, A.P. 500 032

Tel. 27833056 Ref:

Date: 8/3/03

The Secretary,
R. K. Mission Seva Samstha,
Domalguda,
Hyderabad – 009.

Dear Sir,

In the beginning of this month, we read with interest your appeal for funds for the new cancer ward that is being extended in your hospital.

We are aware that your Samstha is doing noble service to the people who need help. Therefore we feel that your Samstha deserves all possible help.

We enclose our cheque for Rs.10,000/- and are sure that your new cancer ward will be of great benefit to many patients.

With best wishes,

We remain,

Yours faithfully,

Sd. XXX
(Umesh Chander)
Secretary
Radha Iron and Steel Industry

It can be observed that letter E is very curt and rude. The tone of the addresser is very arrogant and patronizing. Further, the letter does not follow any of the in-use layout forms. The addresser has also used the archaic business expression 'inst.' The letter is written in a single paragraph. Although the letter is brief, it cannot be called a courteous and a polite letter. On the contrary, Letter F reads well and is very appropriate. The tone of the writer is appreciative and humble. The letter also follows the in-use-indented form of layout. It is clear, brief and can be called a very courteous letter. The letter also conveys understanding and appreciation on the part of its addresser. Hence, secretaries for company correspondence can consider Letter F as a good model of business writing.

Business Report Letter

This letter is a simple 'business report'. It is a blend of the formal and the informal letters: the salutation is generally 'Dear Mr XXX' and the subscription is 'Yours sincerely'. The Regional Manager reports on consumer surveys, company related accidents, insurance claims etc., to the General Manager.

The body of the letter generally has three parts:

1. *The introduction*: the first paragraph of the letter gives a lead to the main part of the report. The introduction to a report contains the background information and the 'why', 'who', and 'what' of the report.
2. *The body of the report*: This contains all the facts or findings. The reporter's remarks are not expressed in this part. If the report requires exhaustive data, an appendix is generally enclosed. The body of the report gives only the gist of the data.
3. *The conclusion*: The conclusion contains the following:

   a. Explication of the facts,
   b. Reporter's remarks, and
   c. Suggestions

Look at a sample business letter report:

TOIPS (INDIA) LTD
Nariman Point, Mumbai 400 011.
Branch Office: 20 R.P. Road, Secunderabad 500 002.

20 May 2002

Mr L. K. Basehwar
General Manager,
TOIPS (INDIA) LTD.
Mumbai.

Dear Mr Basehwar

As decided at the meeting of Regional Managers of the company at Mumbai in January, a survey of customer response to our new product FAIR toilet soap was made in Hyderabad during April 1992.

The survey reflected the mixed response of the customers.

1. 30% of a representative selection of the population use FAIR toilet soap.
2. 60% of the users like the soap, 30% of them find nothing unique in the soap, but would continue using it. 10% expressed dissatisfaction about the soap.
3. Nearly 60% of the users complained that the soap dissolved quickly in water, and lasted for a very short time. However they said that the soap did give enough letter.

In view of the results of this survey, I recommend that the Research and Development wing look into the shortcomings and faults, and their findings be placed before the subsequent meeting of Regional Managers in August 2002.

Yours sincerely,

Sd XXX
(P. K. S. Murthy)
Regional Manager

FORMS OF LAYOUT

Business letters can adopt any one of the many acceptable formats. The acceptable formats are as follows:

1. Indented Form (traditional form)
2. Hanging Indention
3. Block Form (more modern form)
4. Semi Block

These different forms of layout are shown below:

Layout 1: Indented Form

SENDER'S NAME AND ADDRESS

Tel. ________ Ref. ________

E-mail: ________ Date: ________

Mr R. Shermani,
Manager,
Bank of Baroda,
Hyderabad.

Dear Sir,

____

____

____

____

____

____

____

____

Complimentary close
and Signature

Layout 2: Hanging Indentation Form

SENDER'S NAME AND ADDRESS

Tel. __________ Ref. __________

E-mail: __________ Date: __________

Mr R. Shermani,
Manager,
Bank of Baroda,
Hyderabad.

Dear Sir,

____

        ____

        ____

        ____

____

        ____

        ____

        ____

Complimentary close
and Signature

Layout 3: Block Form

SENDER'S NAME AND ADDRESS

Tel. ___________ Ref. ___________

E-mail: ___________ Date: ___________

Mr R. Shermani,
Manager,
Bank of Baroda,
Hyderabad.

Dear Sir,

____

____

____

____

____

____

____

____

Complimentary close
and Signature

Layout 4: Modified Block Form

SENDER'S NAME AND ADDRESS

Tel. __________ Ref. __________

E-mail: __________ Date: __________

Mr R. Shermani,
Manager,
Bank of Baroda,
Hyderabad.

Dear Sir,

____

____

____

____

____

____

____

____

Complimentary close
and Signature

Layout 5: Semi Block Form

SENDER'S NAME AND ADDRESS

Tel. __________ Ref. __________

E-mail: __________ Date: __________

Mr R. Shermani,
Manager,
Bank of Baroda,
Hyderabad.

Dear Sir,

____

____

____

____

____

____

____

____

Complimentary close
and Signature

ACTIVITIES

1. Draft a business report letter on the basis of a consumer survey conducted for your company's product -an intercom telephone system.
2. You agreed to supply intercom telephones to a newly built hotel by a certain date. Write a letter to the hotel manager requesting for extension of time for supplying the equipment.
3. An electronic gadget purchased by your department last month is not functioning properly. Complain to the manufacturer who has given you a guarantee for one year.
4. A social organization contacts your company and seeks donation for their hospital extension. Write a letter, which is brief, clear courteous and considerate, to the organization in your capacity as Secretary of the company, refusing donation.

"I hate quotations. Tell me what you know."

—Ralph Waldo Emerson

TECHNICAL WRITING

Put it before them briefly so they will read it, clearly so they will appreciate it, picturesquely so they will remember it, and above all, accurately so they will be guided by its light.

—Joseph Pulitzer

Introduction

Technical writing is a typical form of writing different from expressive, expository or descriptive writing. Typically, technical writing requires give and take, a dialogue, a follow up, input and action. Most often, it creates action, causes the person at the other end to react or respond. It is also a form of documentation where processes are described, recorded and analyzed. Documentation could also be about the different phases of a product life cycle or even the responses to a certain experiment or exercise. Most often, these are in the forms of reports. It is important to emphasize here that there may not be a single, one and only procedure of reporting or documenting. It is always need, and situation-specific. There are three factors very important for technical writing – purpose, audience and tone. The purpose of writing and the audience very often sets the 'tone' of a piece of writing. If you are writing to someone above you, you are probably requesting or recommending action. But if it is someone below you, you are directing action, instructing. Again, if you are writing a report solely for the purpose of documenting, your language will be different. Similarly, if you are presenting a proposal and trying to convince your colleagues, your tone will have to be persuasive.

Technical writing can be of various kinds. There can be reports or documents such as proposals, product specifications, or quality test results. There can be instructions like user guides, online help, training and user manuals. There can even be business proposals, status reports, customer documentation, and e-mail reports. All these kinds of technical writing have unique formats, but there are general features that are common to all of them.

All writing is aimed at achieving some purpose or the other. But technical writing is very specifically aimed at achieving certain purposes. A good training manual will do exactly what it is intended to do. A well-designed and well-written piece of technical writing has to take into consideration some important factors even before the process of writing begins.

Present to inform, not to impress; if you inform, you will impress.

—Frederick P Brooks

Organization and Language

After deciding why the document is to be written, the objectives it is supposed to achieve and the audience it is meant for, the most important factor you must concentrate on is the organization and the language.

Organization

One very important factor in technical writing is clear and orderly organization. Faulty organization can result in the information being distorted and hinder communication. To write with clarity one has to first have an outline of what one is going to present. There is no specific format for an outline. But it has to be variable and flexible to suit the subject and scope of coverage. To have a good outline you must have:

- A clear and emphatic summary of the subject matter.
- The perspective from which you are looking at it.
- Evidence in support of the thesis.

Some of the other techniques that can be used to make writing more organized are:

- Introduction to the subject matter.
- Headings in large font to express transition to new ideas.
- Bulleted lists to draw attention to the subject matter and make comprehension easy.
- The use of figures and illustrations and the help of diagrams and graphs to give immediate visual representation of what is being said.
- A summary that restates the main idea and emphasizes the perspective.

> *Documentation is the castor oil of programming*
>
> —*Gerald Weinberg*

> *I guess it's better to have too much information than too little. It makes the manufacturer feel good, keeps writers employed, and makes attorneys work a little harder. Besides, it gives me something to read as I'm eating those chewy white peanuts that came with my new computer.*
>
> —*Dave Glardon*

ACTIVITIES

5. Read the following passage. Give it a title. Identify the main idea, the supporting idea and organize them systematically. (Your answer will appear similar to the pattern given below the passage)

Post-independence generations of Indians came to associate the term 'scientific temper' with the maker of modern India, Jawaharlal Nehru. It was he who invited fellow Indians to participate in the adventure of building a strong and forward-looking nation with the help of modern science and technology. A scientific temper required a new approach to life and things, free from obscurantist ideas steeped in irrational dogmas and superstitions. It needed an educational system expressly wedded to the promotion of democratic, socialist and secular values. In actual terms, it meant adoption of the kind of culture that prevailed and prospered in the last three hundred years in Europe and America. The industrial revolution and all that made it possible was part of that culture. Empiricism and experimentation are its backbone. A certain objectivity and avid pursuit of truth based on verifiable facts are its essential concomitants. Such a temper is, in fact, both a cause and an outcome of the modern world.

Title ______________________

1. ______________________________

    i. ______________________

    ii. ______________________

    iii. ______________________

    iv. ______________________

2. ______________________________

    i. ______________________

    ii. ______________________

    iii. ______________________

6. Given below is a write up on the different kinds of crossword puzzles. Give it an appropriate shape, edit it and format it with attention to the font size, numbering pattern, etc.

Three types of crosswords

Prize competition—here the winner surely gets a big prize. It is primarily a gamble where many answers are equally appropriate. By paying a small entry fee here, one can win a big prize.

Intellectual crosswords—here there is only one possible answer to every question. The answer is elusive and calls for some detective work. The clue gives only hints about the word and your comprehension and general knowledge are put to test. It is primarily an intellectual exercise.

Educational crosswords — this is the kind of crossword puzzle where clues or hints about word are given and they have to be found and put in their proper place. The words have to be put in the proper place in the crossword square. This kind of exercise is very helpful as a vocabulary exercise. They can be considered entertaining as simple exercises in problem-solving too!

Theory without practice is sterile, practice without theory is blind.

Technical Description

A technical description can be part of a report or a report itself. It is especially used when the description talks of a new tool, device or a concept. The following are some of the ways in which a technical description is made effective.

Comparisons: technical descriptions often include depiction of unfamiliar objects. This can include description of shapes, angles, colors, size and structures in exact meaningful forms. To do this, we often need to give comparisons with familiar objects. Screws thus have 'heads', saws have 'teeth', switches have 'backs' and the computer has a 'desktop'. Some of the dimensions along which descriptions are often given are: color, texture, size, quantity, shape.

Something that is being described thus could be parallel, perpendicular, cylindrical, grainy, slippery or smooth. Technical descriptions require that the writer should empathize with the reader and have an idea of the visual image that will be created. Thus very much contrary to common belief, a technical writer has to be creative, has to visualize and empathize with the audience to be able to make an effective presentation that reaches out and communicates.

The following is an example of the use of comparison in technical writing:

> A word processor is a piece of software that enables a computer to function like an automated office. With the software the computer works like a typewriter so that you can create your own files: memos, letters, reports and graphics. The computer memory, commonly a disk, acts as a file cabinet from which you can retrieve files, edit them when necessary, and save files infinitely. Acting as a secretary and copier, the word processor controls the format, type font, and the number of copies to be printed on the printer. Some word processors also contain a dictionary and thesaurus to check spelling and word choice and a mail merge function for automatic addressing, which relieves secretaries of these duties.

Generally, a description of an object or device will contain the following:

1. Description of the function; the when and why of its use.
2. The physical appearance of the object and its component parts one by one. This will include a general description of the object or device and include the dimensions, appearance and components.
3. An account of the components in sequential and logical order, including the description of each of them and their relationship with other components. In other words, the manner in which these parts play into the larger function of the devise itself.
4. The definition of the idea or object, its overall characteristics, followed by a detailed description of the parts in a logical order.

If you are trying to talk about the computerization of a hotel management system, for example, you'll have to include the following features:

- A detailed description of why this system has to be computerized. This has to include an account of the existing system; the problems in the system and where and how the changes should be brought about.
- The detailed plans of the proposed system—the requirement, the cost estimation and even the models that can be used.
- The 'designing'—the physical and the logical designing and the "coding"

The last two features that are extremely important are "testing" and maintenance". Part of maintenance can also be preparation of user manuals etc., which need instructional writing and a very exact use of descriptive writing to explain and describe the functioning or products.

ACTIVITIES

7. Describe a computer to an adult rural Indian who has never seen a computer. While doing so, take care to mention the following:
   a. the most general use
   b. comparing it with familiar objects
   c. giving an overall picture of the parts and describing their interconnection
   d. Describing the parts in greater detail, their use etc.

8. Describe a water filter to a person who has seen it but never used it before. Ensure that you give the correct expressions regarding the shape, size, and texture and functioning of the different parts.

Process Description

The technical description of a process talks about how something works. It talks of the overall function of the process and the materials or skills required. The description most often includes a flow chart or a diagram that can illustrate the sequence of action and the decisive points. A typical process description will have the following elements:

- Introduction or definition that says when and why the process is performed.
- The general operation, which gives an idea as to the skills and time required, the pre-operation and the post-operation conditions.
- Description of the steps—why and when it takes place, how long it lasts and what are the human interventions required.

Whether process description is the description of an operation, a mechanism or a system, the following steps have to be followed:

1. **Defining the audience:** Process descriptions are often used to decide whether a new process should be implemented or a project should be started, etc. Careful choice of the language, words and information thus should be effective and relevant to the audience the material is being presented to.

2. **Selecting an organizational principle:** The next important factor in process description is, selecting an organizational principle. The organizational principle for a process is generally chronological. The writer starts with the first action and continues in order till the last. While in some processes, the sequence is obvious, some others require careful examination to determine the most logical sequence. If you were describing the development of cinema, for example, you would clearly follow the chronological order. But if the topic is the complex flow of material from a plant, the sequence has to be based on the audience's knowledge level and the intended use of the description.

> *In the interests of clarity, it seemed necessary to constantly remind myself to pay not the slightest attention to the elegance of the presentation; I adhered conscientiously to the rule of the brilliant theoretician, Ludwig Boltzmann, to leave elegance to tailors and shoemakers.*
>
> *—Albert Einstein*

3. **Choosing appropriate visual aids:** The third important factor in process description is the use of visual aids. Graphic representations like a flow-chart can effectively represent a process and explain the sequences involved. A decision tree can help us to decide whether or not to perform a certain action in a certain situation. Here, at each point, the reader must decide yes or no and then follow the appropriate path.

The following is the description of the hand lever and piston of a paper micrometer, a mechanism used to measure the thickness of paper.

The Hand Lever:

The hand lever, shaped like a handle on a pair of pliers, raises and lowers the piston. It is made of chrome-plated steel and attaches to the frame near the base of the dial. The hand lever is four inches long, ½ inch wide, and ¼ inch thick. When the lever is depressed, the piston moves up, and the hand on the dial rotates. When the hand lever is released and a piece of paper is positioned under the piston, the dial shows the thickness of the paper.

[Analogy]
[Relationship]
[Effect]

The Piston:

The piston moves up and down when the operator depresses and releases the hand lever. This action causes the paper's Thickness to register on the dial. The piston is 3/8 inch in diameter, flat on the bottom and made of metal without a finish. The piston slides in a hole in the frame. The piston can measure the thickness of paper up to 0.300 inch.

[Functioning]
[Description]
[Function]

Activities

9. You have all opened the computer and saved new files. Describe the process to somebody who is operating the computer for the first time.

10. In one paragraph, describe how to start and set your vehicle in motion. Your audience is a group of class of students who are learning about different vehicles. Use this plan:

    a. Name and define the vehicle in the first or the second sentence.
    b. Present the details in several sentences.
    c. Chronologically order the sequence of action.
    d. Roughly try to talk of the link each has with the other.
    e. Make a diagram to show the sequence. If you can, make a flow diagram

> *Good writing is clear thinking made visible.*
>
> *— Bill Wheeler*

E-MAIL WRITING

Basics of E-mail Messages

Electronic mail (e-mail) is the medium of communication that sends and receives messages via particularly designed computer networks. Subsequent to the growth of information technology along with the rapid use of the Internet, e-mail has become the most popular medium of communication. Increasingly, people are using the internet for sending e-mail messages. Undoubtedly, because of its high speed, low cost and efficiency, e-mail today is one of the most vital channels of communication. Similar to business letters and memos, e-mail messages help to strengthen professional and business relations. Routine business dealings and activities would be extremely difficult without e-mail. E-mail can be used both as a means to get in touch with people outside an organization, and to send messages within an organization.

All this entails effective e-mail writing skills as we write a number of e-mail messages everyday. Since e-mails are faster than letters and memos and are used for quick transmission of information and ideas, they serve varied purposes including the following:

- Conveying routine information about new products, procedures, services, policy changes and market strategies;
- Seeking information or resources;
- Inviting people to business meetings, conferences, seminars, workshops, or symposiums;
- Asking for explanations, feedback or clarifications;
- Discussing problems and possible solutions;
- Persuading people to take a course an action; and
- Furnishing feedback, suggestions, or recommendations.

Benefits of E-mail

1. *Speed* is the main benefit of using e-mail. Contrary to regular mail, which may take days or weeks to reach its readers, e-mail reaches almost immediately irrespective of distance, place or person receiving it. We just need to type the name(s) and e-mail address (es) of the addressee and click the mouse on the send button, and our message is sent.
2. *Low cost* is yet another benefit of using e-mail. Because sending an e-mail does not involve printing and copying, it is less expensive than any other channel of communication (that is, postal mail, telephone, fax etc.). A dozen e-mail messages may be sent in twelve minutes and the cost could be as low as six or seven rupees. Interestingly, the size of the message or the distance to the recipient does not affect the cost.
3. E-mail makes for *quick and easy distribution*. Messages can be transmitted to multiple recipients simultaneously without involving repetition and wastage of time.
4. E-mail enables *total flexibility* to the entire process of writing. Before sending an e-mail, the user may edit, revise, modify, and redesign the message without printing or copying it. Furthermore, the user has the option to receive or compose e-mails at one's own convenience.
5. E-mail enables *easy attachment* of files, photographs, clippings, drawings video clips, sound recordings, etc to the message. For instance, resumes, scanned copies of testimonials, transcripts, and other documents can be attached to job applications sent through the e-mail.
6. E-mail is relatively less formal and less structured than letters. It is generally in the form of a private conversation, where the sender wants to say something and expects a response from the reader. Thus, e-mail promotes *ease during upward and downward communication.* While sending an e-mail the writer need not worry about a serious, fixed style of communication and may follow the format of any set pattern or style of writing that suits the writing context.

Features of Effective E-mail Messages

The use of e-mail for business and professional purposes continues to become more wide-based due to its remarkable advantages. It is increasingly becoming the most used professional communication channel. Effective e-mail messages however have certain features, which include conciseness, accuracy, clarity, conversational tone, and single theme.

1. *Conciseness* is the most important feature of an effective e-mail message. An e-mail should contain only the essential information. Unessential explanations, repetitions, wordy expressions, and exaggerations should be avoided. Points should be so organized that the e-mail makes its point with as few words as possible. The recipient may not have the interest or time for a very long and detailed message.
2. *Accuracy* is vital to effective e-mail writing. The format and structure used should be accurate. Correct e-mail addresses should be put in. Messages certainly bounce if an incorrect e-mail address is used. You should check the content of the e-mail for factual accuracy. E-mail messages should be glossed over for spelling, punctuation, and grammatical mistakes.

3. E-mails ought to be *simple and clear*. Simple, familiar, direct and specific words, appropriate linkers, and transitional signals ought to be used to write crisp sentences and paragraphs.
4. The tone of e-mail messages is usually semi-formal but *conversational*. It is advisable to use a tone which gives a personal feel to e-mails. Nevertheless, one should avoid being too informal or emotional. The test is to maintain professionalism without being too official.
5. An effective e-mail message generally deals with only one topic. In order to be focussed, you need to concentrate on a single theme. Developing a single theme logically and substantiating it with related ideas is important.

Format of an E-mail Message

While e-mail systems normally come with a readymade format, we need to adhere to standard writing conventions and implement it effectively. Thus, an awareness of current e-mail conventions and standard practices is very essential. To write an appropriate e-mail, it should be follow a correct format.

When we get an Internet e-mail message, we find that it usually contains many line segments before the beginning of the actual text. These line segments consist of the "header" of the message. Basically, it is a record of the path the message took from the writer's computer to the reader's computer. Headers also contain information of the time and date and indicate if any file is attached to the message.

The three most important items of information in the header are the e-mail addresses of the sender and the recipient, and a subject line that may tell what the message is about. All e-mail messages contain these three items of information.

When a person sends an e-mail message, the programme usually inserts the sender's name, return e-mail address, and date automatically. Therefore, the sender need not type the name, e-mail address, and date again. The sender just needs to fill in the "To" line with the recipient's e-mail address, the "Subject" line with a clear and concise description of the subject of the message, the Cc (carbon copy) line with the e-mail address of anyone who is to receive a copy of the e-mail message, and the Bcc (blind carbon copy) line with the e-mail address of anyone who is to receive a copy of the e-mail message without the knowledge of the main recipient and the Cc recipient. If many names are kept in the Bcc line, each of the recipients will be ignorant about who else is included as Bcc recipient.

An E-mail includes the following:

- Heading
- Salutation
- Body
- Closing
- Signature

Heading:

The heading segment of an e-mail includes the following six items:

- Date
- From
- To

- Subject
- Cc
- Bcc

The Date line indicates the date the e-mail was written. It includes the day, month, year and the exact time. While sending an email message, the date line usually appears automatically.
Example: Date: Tue, 25 Nov 2009 12:58:20 + 0100(BST)

The 'from line' contains the sender's name and e-mail address. The name does not include any personal title such as Ms, Mrs, Mr, or Dr while sending an email message, the return address usually appears automatically.
Example: From: "kishore kumar" kishkum@yahoo.co.in

The 'To line' includes the recipient's email address.
Example: sunita@gmail.com

Subject The subject line summarises the topic of the email in a few words. It includes clear and complete information about the theme of the e-mail in phrase form. Figure 1 shows email heading.
Example: Subject: New syllabus details

Cc The Cc line (carbon copy) may include the e-mail address of any one who is to receive a copy of the e-mail message. It is an optional line.
Example: Cc: mukesh@cal.vsnl.net.in

Bcc The Bcc line (blind carbon copy) may include the email address of anyone who is to receive a blind copy of the email message. That is, it includes anyone who is to receive a copy of the e-mail message without the knowledge of the main recipient and the Cc recipient. It is an optional line.
Example: Bcc: swami_nanda@rediffmail.com

Figure 11.1 shows a sample e-mail heading.

Mail	Addresses	Calender	Notepad		kashyapn@yahoo.co.uk[Signout]	
Checkmail	Compose				Search Mail – Mail Options	
Check Other Mail	Previous	Next	Back to messages		Printable view – Full	
[Edit]	Delete	Reply	Reply All	Forward	As attachment	Move to folder
Folders[Add]	Date: Tue, 25 November 2009 16:2:21 + 0100					
Inbox	From: "Murali C" <cmkri@gmail.com>					
	To: "Nandilal Yadav" <nanav@yahoo.co.uk>					
Draft	Subject: UGC Project					
Sent	Cc: ctsvkksg@rediffmail.com					
Bulk [Empty]	Bcc: sivanishiks@yahoo.com					
Trash[Empty]						
My Folders[Hide]						

Figure 11.1 E-mail heading.

Salutation

A salutation should be used as in Figure 11.2, if e-mail is being used as a means to reach out to people outside the senders organization. The same name as in the To line can be used with a personal title such as Ms, Mrs, Mr, or Dr. However, salutation may be omitted if the e-mail is being used to send information inside the sender's organization.

Examples: Dear Dr.Kumar, Dear Professor Shivram, Hello Jaya, Hi Ram

Mail	Addresses	Calender	Notepad		kashyapn@yahoo.co.uk[Signout]	
Checkmail	Compose				Search Mail – Mail Options	
Check Other Mail	Previous	Next	Back to messages		Printable view – Full	
[Edit]	Delete	Reply	Reply All	Forward	As attachment	Move to folder
Folders[Add]	Date: Tue, 25 November 2009 16:2:21 + 0100					
Inbox	From: "Murali C" <cmkri@gmail.com>					
Draft	To: "Nandilal Yadav" <nanav@yahoo.co.uk>					
	Subject: UGC Project					
Sent	Cc: ctsvkksg@rediffmail.com					
Bulk [Empty]	Bcc: sivanishiks@yahoo.com					
Trash[Empty]	Dear Mr Yadav					
My Folders[Hide]						

Figure 11.2 E-mail salutation

Body

The body of an e-mail message describes, explains, and discusses the main idea of the e-mail. The content of the e-mail should be organized logically. The first paragraph may begin with a warm opening followed by a statement of the main point. The next paragraph should begin by justifying the importance of the main point. In the next few paragraphs, justification should be continued while providing background information and supporting detail. The closing paragraph should reinforce the purpose of the e-mail and, sometimes, request some type of response or action. Figure 11.3 shows the body of an e-mail message.

Closing

An external e-mail message may be concluded with a suitable closing such as Best regards, Kind regards, Regards, Sincerely, Yours faithfully, Thank you and regards, All the best, Best etc.

Signature

The signature line in an e-mail message generally contains only the writer's name. However, it can sometimes include the title and organization of the sender.

Mail	Addresses	Calender	Notepad		kashyapn@yahoo.co.uk[Signout]	
Checkmail	Compose				Search Mail – Mail Options	
Check Other Mail	Previous	Next	Back to messages		Printable view – Full	
[Edit]	Delete	Reply	Reply All	Forward	As attachment	Move to folder

Folders[Add]
Inbox
Draft
Sent
Bulk [Empty]
Trash[Empty]
My Folders[Hide]

Date: Tue, 25 November 2009 16:2:21 + 0100
From: "managingeditor" <managingeditor@enginel.com>
To: "Sunita" <sunitam@yahoo.co.in>
Subject: Engine Communication
Cc: ruthsimon@gmail.com
Bcc: douglasker@yahoo.com

Dear Sunita

The new issue of Engine (Issue No. 8, Nov-Dec 2009) is now released and can be viewed at www.engine.com. The 'focus' is on British-American Bridge models that have much in common and seamlessly merge into each other. Edited by Dr Kumar, the section presents interesting vignettes from both the sides. In the 'Feature' section we present the fascinating design of the short FOBs edited by Alak. It takes a brief look at the short FOB forms in British civil engineering history. The regular sections of Structure dynamics articles, book-reviews, strength of materials add to the rich blend of fare on offer.

There has been a delay of almost a week in the release of this Issue, mainly due to work in the family. We regret this delay and hope that it will only be a one-time occurrence. I am thankful to my colleague Anandi, who gamely took on the onerous task of editing the Issue in my absence. Not being too comfortable with the technical aspects, she literally had to dare through the rough tide. She was ably supported by Anjan from the US and Chandrsekhar from Hyderabad. We are grateful to both of them.

Our next issue – the Anniversary Issue – will take a look at recent trends in Indian Structural Works, under the editorship of Prof G Raj of JNU, Delhi. We invite our members to contribute to the section. These will hereafter become annual features to be presented in the Jan-Feb 2010 issue.

With kind regards

Sravan

(G Sravan Rao, Managing and Chief Editor, Engine)

Figure 11.3 Body of an e-mail message.

Regular E-Mail Practices

As e-mail messages are methodical attempts to work together with colleagues and other professionals, regular e-mail practices n eed to be followed. The following suggestions will help in organizing and presenting e-mail messages methodically:

Follow Established E-mail Principles Every organization maintains certain rules regarding electronic communication. Some organizations may consider certain messages unsuitable for the company e-mail system. In many of the organizations, e-mail is not used to send confidential messages such as confidential reports, company secrets, matters relating to organizational business and sensitive personal business dealings. It is important for us to be conversant with established e-mail conventions of the organization we work in. As a rule, e-mail is not used to send confidential, complex, embarrassing or sensitive messages. As e-mail creates a permanent record that can be used against the sender, it should not be used to convey anything that is inappropriate for public consumption.

Check Mailbox Regularly As speed is the main benefit of using e-mail, everyone wants a quick response to their e-mail. We ought to check our mailbox daily so that we can read every e-mail message sent to us and respond appropriately. In case, we cannot respond because of some reasons, an acknowledgement should be e-mailed back.

Correctness Many people have a tendency to be casual while sending e-mail messages. Care should be taken about accuracy of information as well as of presentation. It is very important that the sender is first satisfied regarding the accuracy of information that is being sent before clicking the 'send' key. The following should be cross-checked.

- the electronic address/addresses of the receiver;
- the subject, Cc, Bcc lines;
- content of the e-mail message; and
- attachments, if any.

Further, it is important to edit and revise e-mail messages so as to improve their presentation quality. E-mail messages should be reviewed to analyze whether they can complete their purpose. They should be edited to correct their format, mechanics, grammar, spelling and punctuation. The spelling and grammar check inbuilt into the programme can be used for this purpose.

Brevity E-mail messages should preferably be kept short. Very few people like reading very lengthy e-mail messages. Unessential information, wordy expressions, repetitions, and exaggerations should be avoided. The e-mail message should make its point in as few words as possible. Sentences and paragraphs should be written short.

Readability In order to make a message reader-friendly, the sender must keep the computer screen in mind while composing the e-mail message. Bring in features such as introductory summary, headings, sub-headings, listings, bullets in order to improve readability if messages are slightly long.

Tone You ought not to use a tactless or negative tone as this can lead to confusion and misunderstanding. A formal but conversational tone, which lends a personal touch to your e-mail is better. The sender must adopt the expressions to suit the context of the message and the needs of the readers. First person pronouns (I, we) and conversational contractions (you'll, he'll, she'll, can't, don't, doesn't) can be used by all means.

Email Writing Process

In order to compose an effective e-mail message, the writer should identify the problem/context that led to the writing of the e-mail message and analyze the possible audience to understand their needs. The writer should decide the scope of the message, and prepare an outline of the main points that are to be included in the e-mail. Once the sender has decided what should be covered in the e-mail, the message can be organized by selecting an appropriate organizational pattern. The sender can then write the first draft. After review and revision of the first draft, the final draft is written.

If a response to an e-mail message has to be made, the message should be read carefully to understand what the writer wants. After deciding the scope of the message and organizing the message, the first draft can be written. The draft is reviewed, revised and edited before composing the final draft.

SUMMARY

- Business letters are letters written in the context of business transactions between individual business personnel or between business organizations.
- Business letters normally follow a general layout.
- A good business letter has to be courteous, considerate, precise and clear, complete, and brief.
- A simple business report is a blend of the formal and the informal letter.
- Business letters can adopt any one of the many acceptable formats. The acceptable formats are the indented form (traditional form), the hanging indention, the block form (more modern form), and the semi-block form.
- Technical writing is a kind of documentation that involves the act of recording a process or an event for future reference.
- The three important factors for technical writing are purpose, audience and tone.
- In technical writing, it is important to define the purpose and define the audience.
- Organization, language and presentation style are equally important.
- Many phrasal verbs and prepositional phrases are used in formal/ technical/ business writing contexts as they have different ranges of use, meaning or collocation.
- Electronic mail (e-mail) is the medium of communication that sends and receives messages via particularly designed computer networks.
- Because of its high speed, low cost and efficiency, e-mail, today, is one of the most vital channels of communication.
- E-mail can be used both as a means to get in touch with people outside an organization, and to send messages within an organization.
- The benefits of e-mail speed are low cost, quick and easy distribution, total flexibility, easy attachment and ease during upward and downward communication.
- Effective e-mail messages however have certain features, which include conciseness, accuracy, clarity, conversational tone, and single theme.
- An E-mail includes heading, salutation, body, closing and signature.
- As e-mail messages are methodical attempts to work together with colleagues and other professionals, regular e-mail practices need to be followed.

REVIEW QUESTIONS

1. Describe the features of a typical business letter.
2. What are the different layout forms of modern business letters?
3. What are the features of a well-written business letter?
4. In technical writing, why is it important to identify the purpose of writing and the audience? Explain the significance of these two.
5. What is the significance of comparison in a technical description?
6. Describe a duster used to clean the writing board in a classroom. Carefully use the priciples of technical description in doing this.
7. What kind of a language is preferred in technical description? Explain with suitable illustrations.
8. Assume that you are Shankar Rao, a student doing B.Tech. in mechanical engineering at JNTU, Hyderabad. Write an e-mail message to Bharadwaj Kumar, the personnel manager of Bluestar Industries (e-mail address: bharadwajk@gmail.com) requesting him to allow you to do short vacation training at the Balanagar plant of the company, as part of your academic assignments.
9. Your are Narasimha Rao, Purchase Officer, BHEL, Hyderabad. Write an e-mail to the Sales Manager of Thambi Office Needs Limited, Secunderabad Office. You want fifty desktop computers at the quoted price of Rs.23, 000/- each. Request the Sales Manager to send details regarding the payment system, delivery charges and the delivery time.

CHAPTER 12

Report Writing, Project and Proposal Writing

Chapter Objectives

This chapter aims to give the students the basics of report writing. It gives the basic structure of reports and provides samples of report types, the letter form and the schematic form of report writing. The specific report forms that have been discussed are the feasibility reports, the empirical research report and the proposal format.

This report, by its very length, defends itself against the risk of being read.

—Winston Churchill

REPORT WRITING: THEORY AND PRACTICE

Modern corporate culture is highly dependent on management information system. This system is based on reporting. Important decisions, facts and information have to be conveyed at various levels and at various stages. The information has to be accurate and up-to-date. The data collected must be properly organized and valid. Ensure that it gives the necessary information suitable for further analysis and action by the management. At this point it is important to remember that there is no single way of writing a report. Many organizations have their own formats. Formats may also vary depending on the kind of report being written and the purpose it is supposed to serve. At times, a report can even be put in the 'memo' format if it is being communicated to someone within the organization. Or it can be a voluminous research report giving factual details and analysis.

Structure of a Detailed Written Report

The structure of the report can be divided into three main parts—The front matter, the main body, and the back matter.

The Front Matter

The front matter consists of the title page, the forwarding letter, the preface, the acknowledgements, table of contents, the abstract and the summary.

Title Page :

The title page of a formal report generally has the following features—
The Title
The Sub-title
Name of the author
Name of the authority for whom the report is written
The contract, project or job number
The date
(Although there is no strict rule as to how the title page should be designed, the following is a sample of how it could look).

Title Page

Report Number: 385

HYDERABAD TEXTILES CORPORATION

A REPORT
ON

INSTALLING A NEW PRODUCTION UNIT

Prepared for

The Managing Director
By K. L. Kumaraswamy
Special Research officer

30th Sep'02

Forwarding Letter and Preface

A forwarding letter is a kind of introductory letter through which the author establishes a rapport with the reader, puts the book or project in the proper perspective. It also makes some important points regarding the scope, contents and purpose of the book. The preface serves more or less the same purpose but while the "preface" is written by the author himself, an expert customarily writes a "foreword". In most cases it might suffice to simply have a preface. It is also possible at times to have a table of contents, an abstract or a summary after the first page and directly go on to the "Introduction ".

Such decisions largely depend on the kind of report you are writing and the usual practice at the place of work.

Acknowledgements:
In the acknowledgements, we generally list out the names of the people and organizations that have helped in the making of the report.

Table of Contents:
If a report is several pages long and deals with various aspects of a problem, it is necessary to have a table of contents. Its function is to give the reader an overall view of the report, help him locate a particular topic or subtopic. Given below is the structure of the table of contents. Note that the Arabic numerals begin only from "Introduction" onwards.

Table of Contents

Abstract / Summary:
Most reports consist of a synopsis, which is called an abstract or sometimes an executive summary. An abstract tells what the report is about and gives the extent of coverage. A summary gives the substance of the report without any illustrations and explanations. However, a summary will give the method of analysis, the significant findings, the important conclusions

and the major recommendations. The abstract is generally about two to three per cent of the original while the summary is about five to ten percent. Often, a combination of both is presented as the synopsis. The executive summary forms an important part of the report. It should contain the following:

- why the work was done
- how the work was done
- what was found—results, conclusions, recommendations.

A good abstract should be complete, concise, specific and self-sufficient. And it is to be written only at the end of a report.

Main Body

Introduction

An introduction familiarizes the reader with the subject of the report. It lists out the following items of information:

- Historical and technical background
- scope of study with its limitations and qualifications
- methods of collecting data and their sources
- authorization for the report and terms of reference; and definitions of special terms and symbols.

While preparing the main body of the report, it is necessary to divide it into the main and the sub points. This improves clarity and aids understanding. An "Introduction" sets the scene and prepares the reader. Thus it needs to state in a sincere and unambiguous manner that which it is going discuss.

Discussion or Description:

The main business of the report is discussed in this section. Containing most of the illustrations, it fills most of the report. The main function of this part is to present data in an organized form, discuss their significance, analysis and the results. If the data is too voluminous, it is given in the appendix. There is no one particular procedure for writing the discussion. Some writers prefer to use a backward order i.e., first stating the results and then describing how they were arrived at. Or at times, the data and procedure is mentioned first and the results follow.

Conclusions:

This term refers to the body of logical inferences drawn and the judgements made on the basis of analysis of data presented in the report. All conclusions are supported by what has gone before; nothing new is included at this stage. If their number is large, they may be itemized in the descending order of their importance.

Recommendations:

Conclusions comprise the inferences and the findings, whereas the function of recommendations is to suggest the future course of action. A busy executive may sometimes read only this part of the report and take decisions. Thus, recommendations are formulated after considering all aspects of the question examined in the report. The terms of reference generally indicate whether recommendations are required or not. If required, recommendations can be listed out in the descending order of importance. If their number is more, they may be grouped under different subheadings.

Back Matter

Appendices:
The appendix contains material, which is necessary to support the main body of the report, but is very lengthy to be included in the text. It should be such that, not reading it should not effect the reader's total understanding of the report. However, it should, if one wishes to examine, provide all the evidence and documents.

Bibliography:
A bibliography is a list of published or unpublished works, which are consulted while preparing a report. While writing a bibliography, care has to be taken about the order of authors or authors' names and surnames, the sequence of details, the punctuation marks, and the layout.

Examples:

1. Atkins, Douglas, G., and Michael L, Johnson. Eds. 1986. *Writing and Reading Differently: Deconstruction and Teaching of Composition and Literature*, Lawrence: University Press of Kansas.
2. Sundarajan Rajeshwari. Ed. 1992. *The Lie of the Land: English Literary Studies in India*, Delhi: OUP.

Glossary:
A glossary is a list of technical words used in the report, with their meanings explained. The need for a glossary depends on how familiar the readers are with the topic discussed.

Index:
The index serves as a quick guide to the material in the report. It helps the reader to locate easily any topic, sub-topic or an aspect of the contents. The index is useful in bulky reports where the table of contents is not sufficient. Entries are made in alphabetical order with cross-references. The page numbers on which information about an entry is available are mentioned against it.

> *Not that the story need be long, but it will take a long while to make it short.*
>
> —*Henry David Thoreau*

Apart from the detailed report above, there can also be the following kinds of reports—

Reports of Routine Nature

In all organizations, there are routine reports that go to management personnel at fixed intervals like every day or every fortnight or every month or every quarter. These are called routine reports. They are related to matters relating to production, labour efficiency, cash flow and sales performance. In fact, these routine reports are mainly statistical, and the clerk or the junior executive in some cases can collect the data in a predetermined form and present it without much use of language. In efficient organizations, there are readymade forms for major reports and only the figures need to be filled in. The function of these routine reports is to collect precise data and submit it on time to enable the management to take corrective action before it is very late.

Reports that Are Special

At times, vital decisions of the management are dependent on special reports. These reports concern things that are not repetitive and routine in nature. When the management calls for a report, it also specifies who has to report and to whom. The instructions further contain a clear-cut statement of the objective behind seeking the report. At times, the special reports are a consequence of routine reporting. For instance, let us say shortage of raw material is reported several times after adequate material has been ordered. The management may suspect thefts of material and may ask a person or a committee to investigate and report. For many reports of this kind, it is necessary to collect data, interpret and analyze it, and even give suggestions. Data can be collected by visiting a place, directly assessing a situation, interviewing, corresponding with people or even holding meetings.

Points to Remember

- A report is a statement of facts
- Clarity and conciseness are of great importance.
- Charts, tables etc can help clarity.
- Passive voice suits a report more than active voice in expression. At appropriate places, it helps to have expressions like:

  "It was observed that..."
  "It is recommended that ..."

METHODS OF REPORTING

There are two methods by which individuals can write reports:

The Letter Method

It may take the form of an ordinary letter. That means that usual formality of heading, address, inside addresses, salutation etc. should be maintained in the report. Below the salutation, the title of the report should come.

The next paragraph should show what procedure the report writer followed and what persons he interviewed. The rest of the paragraphs should show the findings and the last paragraph should give the conclusions or the considered opinion of the report writer. In short, it is just a more systematically written letter.

The Schematic Method

This method is more common in committee reports.

Here, the title of the report is given first. Then, the contents of the report are given under the following subtitles.

a. *Terms of reference*: This section should give the source or the authority that ordered the report. It also describes the scope of the report and its objectives.
b. *Procedure*: This section should give the method followed by the report-writer. It may mention the places visited, persons interviewed, etc.
c. *Findings*: The facts as they exist are presented in proper order here.
d. *Recommendations or Conclusion*: This section should give the considered opinion of the report–writer based on the facts as interpreted by him. It has got to be an expert opinion based on incontrovertible facts.

The end may have the signature and designation of the report writer

> *Easy reading is damn hard writing.*
>
> —*Nathaniel Hawthorne*

The following is a report presented in both the letter method and the schematic method:

1. Report on workers' discontent presented in the Letter form

DECCAN ENGINEERING CO. LTD.
2-2-344, Balanagar, HYDERABAD 500 011.
Ph.040-27938966/67/79

April 24, 2003

The Directors,
Deccan Engineering Co. Ltd.
Balanagar,
HYDERABAD 500 011.

Report on workers' discontent at the Company's auxiliary unit

In accordance with the instruction given to me on April 9, 2003, I visited the auxiliary unit of the Company in order to find out the cause and extent of discontent among the workers.

I interviewed supervisors, plant-operators and 25 workers selected at random in this context.

I found out the following facts that are responsible for this trouble:

There seems to be widespread discontent amongst workers. Several other personnel pointed out that the number of incidents of breach of discipline and general non-cooperative attitude had gone up considerably. They expressed the view that the situation was deteriorating and that a serious outbreak of misbehaviour might result. The workers, on their part also had some complaints. According to them the supervisors were inadequate; the workers were given long hours on the production line and skilled personnel were not eager to explain the working of the machinery to apprentices. There were other avoidable lapses that the workers resented. The skilled personnel, to save their own time, leave the machines dirty so that the unskilled workers have to clean them. Safety precautions in the machine room are continually being ignored. There is a widespread belief that the rates paid to workers are the lowest in the area.

I am of the opinion that immediate action must be taken to prevent further disturbances. It is also felt that there is evidence that all the complaints are well-founded. In particulcuar, ignoring safety precautions is an offence against the Factories Act and must be stopped.

Regarding pay and allowances, it is necessary to bring to the notice of the workers that our rates compare favourably with those paid to apprentices in other similar units. It is therefore my considered opinion that the following suggestions will go a long way in reducing the discontent:

- A special officer may be appointed and he should be asked to draw up a systematic program for the unit.
- All supervisors must be strictly advised to keep machinery clean and observe all safety precautions.
- The Personnel Officer must point out to the workers that the rates of payment enjoyed in this factory are very favorable.

Yours faithfully,

Sd/-

Trivikrama Rao

Company Secretary

2. Report on workers' discontent in the Schematic form

The same report can be presented in the schematic form with sub-headings for different points.

DECCAN ENGINEERING CO. LTD.

2-2-344, Balanagar, HYDERABAD 500 011.

Ph. 040-27938966/67/79

Report on workers' discontent at Company's Auxiliary Unit

Terms of reference: In accordance with the instructions given to me on April 9, 2003, I visited the auxiliary unit of the Company in order to find out the causes and extent of discontent among the workers.

Procedure: In this connection, I interviewed supervisors, plant-operators and 25 workers selected at random.

I found out the following facts that are responsible for this unrest.

Findings: There seems to be widespread discontent amongst workers. Several other personnel pointed out that the number of incidents of breach of discipline and general non-cooperative attitude had gone up considerably. They expressed the view that the situation was deteriorating and that serious outbreaks of misbehaviour might result.

The workers, on their part, also had some complaints. According to them the supervisors were inadequate, the workers were given long hours on the production line and skilled personnel were not eager to explain the working of the machinery to workers. There were other avoidable lapses that the workers resent. The skilled personnel, to save their own time, leave the machines dirty so that the unskilled workers have to clean them. Safety precautions in the machine room are continually being ignored. Further, there is a wide spread belief that the rates paid to workers are the lowest in the area.

Conclusion: I am of the opinion that immediate action must be taken to prevent further discontent and ill will. There is evidence to believe that all complaints are well-founded. Deliberately ignoring safety precautions is an offence against the Factories Act and must be stopped.

Regarding pay and allowances, it is necessary to bring to the notice of the workers that our rates compare favourably with those paid to apprentices in other similar units.

Recommendations: It is therefore my considered opinion that the following suggestions will go a long way in reducing the discontent:

- A special officer may be appointed and he should be asked to draw up a systematic program for the unit.
- All supervisors must be strictly advised to keep machinery clean and observe all safety precautions.
- The Personnel Officer must point out to the workers that the rates of payment enjoyed in this factory are very favorable.

Sd/-

Trivikrama Rao

April 24, 2003 Company Secretary

Feasibility Report

A feasibility report is one of the most frequently used report forms. It is generally written as an evaluation of whether or not a particular course of action is desirable. It is written to help the decision-makers choose between two or more courses of action. Sometimes, the choice is between maintaining the status-quo and choosing the alternative being suggested. Sometimes, the choice can even be between two or more choices when the decision about the change has already been taken. An automobile industry, for example, might be considering the possibility of using hard plastic for certain parts for which metal has always been used. The choice here could also be between hard-plastic or some other alloy recently discovered that is costly but more durable.

The feasibility report has to make a very careful consideration of :

- The various alternatives or the alternative suggested.
- The methods used to test the viability of using the alternatives.
- The advantages and disadvantages of the various alternatives. (This can be presented in a tabular form to make comparison easy at a single glance).
- The alternative that you think is viable and why.

Structure of the Feasibility Report

The following are the main parts of a feasibility report. Remember that only the main parts are being mentioned here. Other elements of the report such as the title page and synopsis have to be written according to the conventions of the workplace and the kind of report being written.

The Introduction: "Introduction" of the feasibility report is generally the answer to the question: why should we consider these alternatives? For this, one has to identify the problem the feasibility report will help us solve, study the alternative courses of action considered and also briefly consider the methods used to investigate and arrive at the conclusions.

Evaluation: In a way, the evaluation is the heart of the feasibility report. A detailed point-by-point evaluation of the course of action being recommended is what the feasibility report is about. The manner in which we present the matter for evaluation will determine the type of report that is being prepared. An important part of "evaluation" is dismissing unsuitable alternatives.

Conclusions: The conclusion will consist of a detailed assessment of the course of action the evaluation section has spoken about. You can mention your conclusion in more than one place in the report. The summary at the beginning can mention the conclusion. The introduction, again, can mention the conclusion finally arrived at. And finally, you should be able to provide a detailed description of the conclusion in the section following the "Evaluation".

Recommendation: A feasibility report generally ends with what the writer of the report feels about the alternatives. Since you have studied the situation so thoroughly, you should be in a position to make sensible recommendations that suggest a course of action.

Points to Remember

- A feasibility report is written to help people choose among alternatives. Present the alternatives carefully.
- State the methods you have used for investigation.
- Decide how to present the matter for evaluation after carefully considering what the reader needs from the report.
- Always put the most important point first.
- Mention your conclusion more than once: For example, in the summary, the introduction, etc.
- Make firm and sensible recommendations. After all, you have studied the problem minutely.

Activity

1. You are an engineer in the army responsible for laying roads to remote regions. There is an army post in the remote area of Gangtok very near to the border, which depends only on air traffic for its connectivity with the rest of the country. In view of the increased communal problems and terrorism, there has been a proposal to lay a road through the mountainous region and make the place more accessible. The task however is formidable and will require a lot of finance. There has been very mixed opinions regarding the road at the decision making level. You have been given the task of consulting other experts, studying the situation and giving feasibility report on the issue.

The Empirical Research Report

The empirical research report is prepared after careful investigation or observation. It is prepared when, for example, scientists are trying out a new technique to improve the communication system through satellites or a group of social scientists are trying to find methods to improve the creative faculty in students. This kind of report will have the following structure.

Introduction	The subject and objectives and the research The importance Methods of collecting data
Results / Facts	The results of the research
Discussion	Interpretation of the results
Conclusions	The significance of the results
Recommendations	The action to be taken next

Introduction: The Introduction of a research report should begin with the announcing of the subject. If you have been working on genetic engineering for example, to produce a superior variety of rice, your introduction can begin like this.

This work is primarily concerned with the genetic engineering techniques that can be issued to enhance the quality of rice grains being produced in our country.

Importance of the work: This work is extremely important keeping in mind the fact that India has to compete now with the global market with respect to food products.

Objectives of the Research: A research report has to have carefully defined objectives. This helps the writers of the report to be focused, select the appropriate research methods and shape the manner in which the result is to be interpreted. The research on the grain quality, for example, can have the following approach.

Particularly, the work shall look into how better grains can be engineered and supplied to the farmers at less cost. We also would like to look into the possibility of making them longer and if possible, aromatic.

Our aim is to have it ready by the time the Indian farmer is made to face the global challenge.

The Method: Details about the method of research help in many ways: It convinces the reader about the soundness of your research proposal, it gives the reader a rough idea of what your research result will be like. And finally, it becomes a documented guideline of the steps that need to be taken if similar research is to be conducted. The methods of the report discussed earlier could read like this:

First, the grains widely used by farmers were taken up for analysis and compared with some of the best quality grains in the global market, etc.

The Results: The results of the study undertaken should be presented in a form that is easily accessible to the readers. It could be put in the form of a table and explained diagrammatically, if necessary.

Discussion: The 'Discussion' will include primarily the interpretation of the results. It is possible to sometimes present the *Results* and the *Discussion* together. Even if they are presented in separate columns, it is necessary to weave them together. Sometimes, if there are many aspects of the *Results* that are significant, every aspect has to be taken up individually and discussed at length.

Conclusions: After interpreting the results, it is important to explain their significance in the light of the original stance taken in the "objectives". It is necessary to restate the results and closely coordinate it with the objectives. In a way, it should tie up the loose ends, restate the focus and make the study a coherent whole. in the report being discussed, the conclusion could read like this.

The study shows that the quality of the grains can be enhanced. The process will necessitate sophisticated technicalities and a few imported machinery. As for the cost factor, the initial cost will be more that what the farmers incur right now. But subsequently, the profits will more than make up for the initial expenditure.

Recommendations: If the research project is based on a practical problem, the readers would definitely like to know what the writer thinks should be done. In the example discussed above, the recommendations could be regarding the following:

A number of research stations could be set up to fasten the process, the method of quickly procuring the machinery involved and a loan system could be considered to enable the farmers to make the initial investment.

Points to Remember

- The subject and objective of the research has to be mentioned clearly at the beginning.
- It is important to mention why the work is important.
- Link up the work with organizational goal. Put it in perspective.
- Give details of the method of research. It is important for future research.
- The results should be presented clearly and systematically.
- During the *discussion* interpret the *results* side by side.

- The conclusions should back and address the objectives of writing the report.
- Tie up all the loose ends of the report while making the *Conclusion*. Show how the results relate to the objectives.
- Make sure that the *recommendations* are practical. Take an overall perspective of the situation before framing them.

Sample Report: Empirical Research Report

Report Title

Defining the Shipping Distribution Environment with Electronic Data

Report has been prepared by
Jagdish Mathur

Prepared for:
P. Gopichand (MD)
Marketing division.

December 19, 2001

Table of Contents

Introduction

Presently many companies are using mechanical data recorders to define the shipping environment. However, a few companies do not consider the use of data recorders at all. This paper explains why companies must seriously consider using electronic shock and vibration data recorders.

With a view to showing why the electronic data recorders are better than mechanical data recorders, this report will discuss the following:

- Differences in the two types of recorders vis-à-vis accuracy and functionality
- The possibility of creating better test plans using an electronic data recorder
- How electronic data recorders reduce costs by defining the distribution environment and refuting damage claims?
- The advantages in the facility of downloading and analyzing data by means of an electronic data recorder.

Data Recorders Accuracy

The earlier mechanical data recorders are not as accurate and dependable as the electronic data recorders. Most of the older devices were purely mechanical in nature and used a paper graph or a visual indicator to quantify shock levels. These devices are popular and presently in use due to the relatively low cost and ease of use. These devices may be cheaper, but they do not seem dependable. The accuracy and functioning of the electronic data recorders will be discussed next to show the relative advantage.

Accuracy and Functioning of Mechanical Data Recorders

These devices are of three main types, each having different functions. All these types, however, have limitations. The mechanical data recorders discussed in this paper include

a. One-use impact indicators
b. Shock level recorders
c. Threshold indicators

One-use impact indicators: Many mechanical shock devices are designed for single-time use. These devices detect a problem only if the product has been tilted or tipped beyond a preset limit. They indicate that the shock has exceeded a certain *g* level, indicating the force of gravity. However they do not show just how many *g* levels have been experienced. They also do not indicate how many times the package was dropped.

Shock level recorders: Shock level recorders print shock level data on the chart paper. The height of the spike on the paper indicates the level of shock. These shock recorders are easy to use, but the results are complex requiring thorough interpretation by experts in comparison with electronic PC-based devices.

Threshold indicators: Threshold indicators show those shock events that go beyond a certain level. These devices indicate that the shock exceeded a certain *g* level. However they do not show how many *g*'s the package experienced, nor do they give an idea of how many drops the package received.

All things considered, it is a fact that can be convincingly established that the mechanical data recorders are not as accurate as the electronic data recorders. If the main purpose of using these recorders is to accurately define the distribution environment, then they must be able to furnish reliable data. If the recorders don't furnish accurate data, they are a waste of time and money. Hence, using electronic data recorders, instead of mechanical data recorders, is preferable.

Functioning of Electronic Data Recorders

The electronic data recorders generally make use of a tri-axial accelerometer to catch dynamic events and store them in memory for processing on a PC at a later time. A tri-axial accelerometer can measure shock emanating from all and any direction, as against many mechanical data recorders that can measure shock in only one or two directions. Most of these data recorders are equipped with software programs for downloading the data and converting it into charts and graphs.

The electronic devices show the acceleration/shock that a package receives in actuality. The orientation of the shock is recorded too. Using a mechanical recorder that could read shock in only one direction, such as flat-on-bottom shocks, one realizes how much packaging protection is needed for the bottom of the shipped product. However one would not know how much protection is needed if the package is dropped on its side or the other end.

The orientation of the shock information is as useful as the drop height information. By using electronic data recorders, that are able to measure shock in all directions, one can accurately define how much protection is required for each side of the product. Instead of a package with two inches of foam on all sides, one can make a package with two inches of foam on the bottom, one-half inch of foam on top, and one inch of foam on the sides and ends. Such a design surely saves packaging material, saves money, and, most importantly, protects the product. The expense involved in the purchase of these data recorders is naturally much more than the mechanical models, but the units are more accurate, dependable and valuable.

Accuracy of Electronic Data Recorders

A study undertaken by me (1998) compared two electronic data recorders, the EIS by 'Delhi Instruments'and the FES by the 'Mitra Sensor Technology'. The FES measured shock values with a margin of error of approximately 1% overall, while the margin of error for the EIS varied between 10% and 25% for similar shock values. A mean error of approximately 5% is acceptable by all standards for most shock measuring instrumentation. As you can see in Table 1, the electronic data recorders are more accurate than the mechanical recorders.

Table 1: Data Recorder Accuracy

Recorder type	**% error**
Single use Mechanical	35
Shock level recorder Mechanical	175
EIS Electronic	10–25
FES Electronic	1

Test Plan Development

After working in a test laboratory for a rigorous period, I realized that test planning was one of the most difficult tasks of my job. It was difficult to determine what drop height to use and what vibration intensity to use. Some companies have a tendency to follow blindly the Standards generally used to qualify a product. According to Kalyan Kumar, the Packaging Incharge at the NII of Vishakhapatnam, the test levels in the Standards are generally not high enough.

In other words, if you want to ensure zero product damage, you must use the electronic data recorders to measure the shipping environment, instead of relying only on the Standards.

For a package to be eligible for shipment according to the Standards, it must initially pass a multiple drop sequence of specific heights and a vibration test. The drop heights and vibration test intensities are determined by the test technician, who decides test levels based on the package weight, cost of the product, method of shipment, and the proposed mode of shipment for the package. Stating theoretically, the higher the weight of the product, the lower the drop height it will be subjected to.

The test technician takes into account this type of information when deciding what test intensity to use:

- If the product is heavy, the cost high, and the product to be shipped in a small parcel environment, the technician will very likely choose Assurance Level I (the highest test level).
- However, if the package is heavy, the product is inexpensive, and the product will be shipped by trucking, the technician will probably choose Assurance Level III (the lowest test level).

Standard test levels can underestimate actual shipping conditions. Many people think that by opting Assurance Level 1 they are being conservative. But actually they are not; they are underestimating the actual vibration intensity, this means that your package could pass the Standard test, but when you ship the product it may get damaged.

During the research, test levels were determined by the Standards by analyzing past shipping data and taking the average of these shipping situations to develop standard test levels that try to match these situations. How does one know if a shipping situation is average? One may answer this question by using the electronic data recorders to measure one's shipping environment, and also by defining the actual shipping conditions. Then one can test at those levels to see if one's package will protect the product adequately.

Standards state that if more detailed information is available on the transportation environment, it is recommended that the procedure be modified to use such information. Electronic data recorders are conducive for gathering this kind of information. By determining the real shipping conditions, the time spent to develop a test plan is reduced and, as we all know, time is both money and resource.

We have determined that an important condition that must be considered in a package design is the drop height. In an experiment done by Anant, a small 3-kg cushioned package was shipped 120 kms via a carrier who kept it five days. The package received seven drops with an average drop height of 12.7", the highest drop being 43.8". But the initial results suggested a design drop height of 43". The author stated that this was only an initial test, and that it should be repeated several times to generate a database. If you compare this drop height to drop heights given in Standards you will see a major difference. Standards suggests a drop height of 16", plus one drop of 31". In fact, JTP 18/4 recommends 32" drop.

Reduced Cost by Defining the Distribution

When a package is shipped, irrespective of whether it contains an expensive piece of equipment or perishable food items, shock and vibration can badly affect the contents. There are many benefits that are gained by using the electronic data recorders to rebut damage claims or to define the shipping environment. This section will highlight these benefits. All the benefits add up to cost savings.

Given below is a detailed discussion of the advantages in using an electronic data recorder.

Decreased damage

Decreased damage may be the main reason for using electronic data recorders. Freight damage claims cost businesses billions of rupees a year. You may have the best-designed product, but if it is not packaged properly to withstand the wear and tear of shipping, you could end up with a broken product. Trucking forms the greatest risk of product damage. A package will be handled 10–12 times and will travel on seven different trucks for an average shipment.

The way to cut the shipping damage is to define the distribution environment. A package could be shipped by many different modes of transportation. The hazards for each mode of transportation and for each specific carrier can vary greatly. Although packaging may protect a product in one situation, it may not protect it in another because of transport quality and varying road conditions.

Source Reduction

Another way to cut costs is to reduce waste by ensuring no over-packaging. Over-packaging accounts for big percentage of the money spent on packaging materials. Clearly, by decreasing the amount of material used to package a product, cost is decreased. Further, if the package size is decreased the shipping cost is decreased, since it may weigh less and more packages may fit in a truck. Less material translates directly into money saving on material and shipping costs.

Conclusion

This report elaborates the reasons why companies should consider using electronic shock and vibration data recorders. The older mechanical data recorders were compared to the newer electronic models and it has been proved beyond doubt that the older models used traditionally are not as accurate as the electronic versions. The gains of using the electronic data recorders are many. Some of the factors that have been discussed in detail in the report are ease of developing test plans, cost savings and ease of data analysis.

Both the above two kinds of reports i.e., feasibility and empirical, along with the ensuing section on proposal writing use a certain structure, style and format, which together constitute the dynamics of Project Writing. However the themes of project writing are numerous and diverse. These themes can be projects dealing with empirical studies, feasibility studies and project proposal writing.

PROPOSAL WRITING

A proposal is a persuasive communication, generally written when the readers are to be persuaded to adopt a course of action you would like them to. This is something you will have to often do in your career. You might have to write a proposal for a new product you want to develop or try to persuade your employers to make a few costly changes. Or, may be, you would want to suggest a few changes in the marketing policy of your company. To do all this, you will have to write good proposals, present them persuasively and convincingly. In writing a good proposal, there are basically three things you need to identify.

- *The Problem*: In a proposal, it is important to state the problem convincingly, talk about the needs, and why the proposal is important.
- *Solution*: The solution to the problem is another important feature in a proposal. The readers will be interested in proving what you want to propose exactly and how it related to the problem presented earlier.
- *Costs*: The next most important thing in a proposal is the cost factor. What will be the cost of the changes and is the cost worth the change.

The Overall Structure

As mentioned before a proposal is a highly persuasive communication that has to be carefully. A typical proposal has the following features.

Topic	Point of Persuation
1. Introduction	What do you propose to do.
2. Problem	Why is it important.
3. Objectives	The features necessary to make a successful solution
4. Solution	The detailed, overall plan of what you plan to do.
5. Plan of action Method Resources Schedule Management plan	The resources with which you have to do the work, the schedule you have fixed for it and your plan to execute it.
6. Costs	How reasonable is the cost.

Introduction: At the beginning of the proposal it is important to announce what the proposal is about. If the proposal is complex, you might have to give some background information to help readers understand what you have in your mind. If you have a proposal, for example, to introduce a device or a new software that will help automate the examination system, you can start the "introduction" in the following way:

This is a proposal to automate the examination system in colleges and universities…

Problem: Once you have made clear what you are proposing, it is important to state what problem it will address. You will have to persuade your readers that the problem is significant to them. The manner in which you describe and present the problem is very important for the success of the proposal. Although you show that the project will achieve its objectives and you are equipped to carry it to its logical end, the readers have to be convinced first that the work needs to be taken up. And to do this you will have to present the problem from their point of view and convince them that it will be their concern too. The fixing of the problem can be of different types. One, where the problem is already defined. Second, where the broad area is vaguely mentioned and you will have to hit upon a issue. The third is where you will have to persuade your readers that a problem exists and it needs to be solved.

Objectives: Stating the objectives clearly is a vital link between the problem and the action. It talks about how the action will solve the problem to make the objectives convincing. Be careful and ensure that every objective grows out of the problem. If we take the automation of the examination system, for example, the objective can be to

1. *devise a system where the probability of subjective and careless marking can be minimized.*
2. *make the whole examination process less cumbersome and error-free.*

Similarly, if we take the example of the booking of conference rooms, our objectives can be

1. *To maintain an up-to-date schedule of reservation that will be accessible to every person in the company.*
2. *Enable only the assigned people in every department to make or cancel reservations through their terminal.*

Solution: This is a section that will describe the plans for achieving the objectives. In both these cases, for example, we should be able to give a basic explanation of the computer programming and an idea of how it would help to achieve the objectives. A good solution has the following properties.

1. It addresses each of the objectives.
2. It is desirable. In fact, it is the best way of achieving the objectives.

It is important also to take the reader into consideration when you are presenting the solution. It can be technical, or you might have to simplify it at times if you feel that it will be more acceptable. You may even have to skip some explanation if you feel that they are obvious to the reader.

Plan of action: It is necessary to provide the plan of action because the audience sometimes needs to be assured that you have the planning and the capacity to do what you want to do. Here, you have to give a detailed account of the following:

Resources: It is necessary to be able to tell your audience that you have the resources, the equipments, the expertise etc., necessary to carry out the proposal. At this point it is also necessary to state what else you would need for the work and how they can be procured.

Schedule: Schedule is one of the important elements in a proposal. You should be able to give your audience a very clear picture of how much time the work will take. This is necessary for several reasons.

1. You should be able to say how much of your time you would like to divert for this project.
2. The audience need to feel that the schedule is sound enough for the proposal to be effective. A good proposal that takes too much time might lose its effectiveness. Similarly, an ill-timed proposal may not have the kind of benefits it would otherwise have had.

A proper schedule and the right timing thus are among the most important things in the making of an effective proposal.

Costs: Invariably, a proposal means that you are asking your audience to invest their money. To convince them to do so, you should be able to give a clear picture of the costs to be incurred. It is important to be able to show that the costs are reasonable. This can be done better if you could also give a calculation of how much can be saved as a result of the project.

Points to Remember

- A proposal is a persuasive communication.
- State the problem convincingly.
- Make your objectives clear.
- Link up your solution to the objectives.
- Mention the resources you have at your disposal.
- Present a viable and suitable schedule.
- Be realistic about the cost but show that it is necessary.
- Think from the point of view of the readers. Keep their goals and perspectives in mind when you are trying to persuade.

Activity

2. You are a software engineer who has newly joined a company. The company too has been established only recently. You would like to make a proposal of buying some costly softwares that can enable the company to take up assignments concerning building architecture. You are an expert in the field and you are convinced that the demand will persist and the company can more than make up for the expenditure made in less than six months. You did speak about it to a board member but he felt that the expenditure might be a little too much at the moment. Still, he wanted you to present it at the board meeting. Consider all the reasons you would like to give and make a convincing proposal for the purchases.

SUMMARY

- Reports form an important part of information management in the corporate and industrial segments.
- The report can primarily be divided into the Front Matter, the Main Body and the Back Matter.
- Routine reports are those that go to management personnel at fixed intervals but there are special reports that determine a lot of management decisions.
- A feasibility report is generally written as an evaluation of whether or not a particular course of action is desirable.
- The empirical research report is prepared after careful investigation or observation of an issue or problem
- Project writing involves writing about themes that can be taken up as project work. These themes can be projects dealing with empirical studies, feasibility studies and project proposal writing.
- A proposal is a persuasive communication, generally written when the readers are to be persuaded to adopt a course of action you would like them to.
- In writing a good proposal, there are three things you need to identify. They are, the problem, the solution, the costs.

REVIEW QUESTIONS

1. How would you define a report? Give an account of the structure of a report showing how each of the parts plays a vital role in constructing it.
2. What is the difference between the empirical and the feasibility report? Illustrate with examples.
3. How and why is a proposal important in professional and academic contexts? Give the main elements a proposal should contain.

APPENDIX

Appendix A

Vocabulary Expansion

THE IMPORTANCE OF VOCABULARY

Though symbols, signs, and gestures are used significantly for human communication, it is basically through words that we think and convey our feelings, ideas, and needs to others. Our understanding of the thoughts of others is largely done through words. This ability to understand and communicate depends to a large extent on the size and accuracy of a person's vocabulary, i.e., the total number of words known and used by the person. In order to speak and write effectively, it is necessary to acquire sufficient and diversified vocabulary. This can be done through extensive listening and reading but this process requires much time before it produces appreciable results. Hence a systematic study is needed for enlarging and enriching one's vocabulary. In this appendix, you can study some aspects relating to English vocabulary, which would enable you to express yourself effectively and precisely in speech and writing.

English Vocabulary

Of all the languages in the world, English has the richest and most extensive vocabulary – nearly a million words. Half of them are listed in the Oxford English Dictionary; the other half remain uncatalogued. In comparison, the French language has about 1,00,000 words and German about 1,85,000. No one can ever hope to learn the million words in English nor is it necessary to learn them all as most of them are superfluous to the normal requirements of an ordinary person. The size of a person's vocabulary is determined by the nature of his work and interests. Thus, a scholar will need more vocabulary than a manual worker. It is desirable, however, for everyone to have a vocabulary in the range of 20,000 to 30,000 words, consisting of both receptive and productive categories.

Recognition and Production Vocabulary

All persons have two categories of vocabulary–words they can recognize while listening and reading and words which they can actually use in speaking and writing. The first category is known as Recognition or Passive Vocabulary and the second as Production or Active Vocabulary.

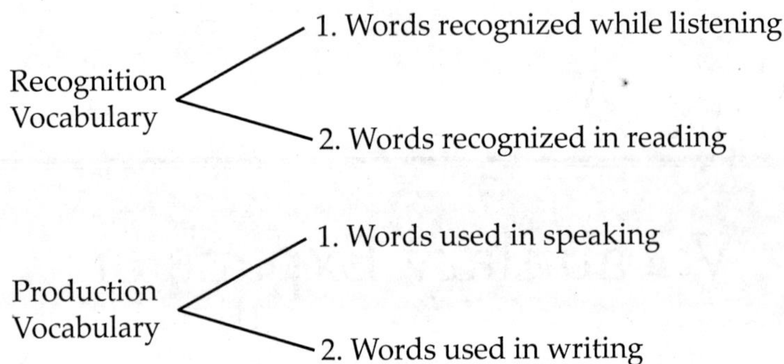

A person's recognition vocabulary is words he she can more or less understand when he/she hears them, or reads them without being sure of their meaning. Production vocabulary is words a person can actually use in speech and writing without doubt or hesitation. The two, however, will not remain constant. Words will move from recognition to production vocabulary when their spelling, pronunciation, meaning, grammar and usage are mastered and when they are used in speech and writing. If neglected, they would be forgotten or remain passive.

Content and Function Words

Words in English can also be classified into two groups–content and function words. Content words are those that express meanings like names (nouns), actions (full verbs) and qualities (adjectives and adverbs). They are unlimited in number, which goes on increasing with time.

In contrast, function words are almost empty of meaning individually. They are grammatical or structural Words like articles, prepositions, conjunctions, pronouns and auxiliary verbs which are required for grammatical construction of phrases, clauses and sentences. They are limited in number (about 160).

Confusibles

There are many English words that are somewhat similar in sound and/or spelling though they are different in meaning. This often leads to confusion and misuse. So, whenever you are in doubt, you must look them up in a dictionary to be quite clear about the differences in their use. Words which often cause confusion can be divided into the following groups.

a. Homophones

Words that sound alike but have different spellings and meanings, are a common source of confusion. Below is listed a selection of such pairs of words.

allowed	aloud	elusive	illusive
altar	alter	fair	fare

ascent	assent	flour	flower
bare	bear	forth	fourth
berth	birth	grate	great
boar	bore	hall	haul
board	bored	hear	here
buy	by	hole	whole
canon	cannon	horde	hoard
cite, site	sight	lead	led
complement	compliment	lessen	lesson
course	coarse	meat	meet
cue	queue	naval	navel
die	dye	oral	aural
draft	draught	peace	piece
dual	duel		
pedal	peddle	tear	tier
plain	plane	there, they're	their
pray	prey	throne	thrown
principal	principle	tied	tide
rain	rein, reign	to, two	too
read	red	its	it's
right, write	rite, wright	know	no
role	roll	waist	waste
root	route	wait	weight
sole	soul	weather	whether
some	sum	week	weak
stationary	stationery		

b. Homographs

Homographs are words that have the same written form but different pronunciations and meanings.

Examples: lead—lead; tear—tear

c. Homonyms

Homonyms are words that are both spelt and pronounced alike but have different meaning.

Example: bear, to bear

d. Confusing pairs of words

accept	except
adapt	adopt
affect	effect
alternate	alternative
angel	angle
breath	breathe
childish	childlike
continual	continuous
council	counsel
dairy	diary
economic	economical
effect	affect

forego	forgo
judicial	judicious
later	latter
loose	lose
meter	metre
negligent	negligible
official	officious
personal	personnel
practical	practicable
quiet	quite
responsible	respectful
virtuous	virtual

e. Collocation

Collocation deals with "what usually goes with what" i.e., an arrangement of words which sounds very acceptable. Faulty collocation leads to comic effect, as for example, if someone were to say, "I want a cup of powerful coffee" instead of "I want a cup of strong coffee". In English, the words "strong" and "coffee" collocate, i.e., usually go together. Similarly, it would be wrong, to say, "I don't want to speak a lie" instead of "I don't want to tell a lie" or "He is a far relation of mine" instead of "He's a distant relation of mine". You can find out about correct collocation of English words by referring to any ELT dictionary.

Phrasal Verbs

Phrasal verbs are idiomatic verb forms made up of a verb and an adverb particle i.e., they are made up of two words or sometimes three words. These verb-adverb combinations have meanings which cannot be built up from the meanings of the individual verb and adverb. Phrasal verbs thus are verbs plus prepositions or adverb complexes that acquire idiomatic meanings. A new verb with a different meaning is produced by a phrasal verb. For example, the expression *go with* when used as a phrasal verb means *to take the same view as* and not *accompany*.

Sometimes the meaning of a phrasal verb can be similar to the original verb as in the following. Example: The train *slowed down* very much before reaching the station (speed has become slow).

Look at other examples

1. I *get up* at 6 in the morning. (wake up)
2. The plane *took off* at 12 in the afternoon. (lifted itself in air)
3. We shall *pick* you *up* outside the bus stand. (receive someone with a vehicle to go back)
4. Arun wants to *give up* smoking at last. (stop a long habit or practice)
5. I can't *go with* you on that point. (agree with)

Phrasal verbs can also have more than one idiomatic meaning:

Example:
Sambar *goes with* (suits) idlis very well.

English is very rich in phrasal verb usage. Phrasal verbs can be learnt easily with the help of a dictionary and also by thinking out situations in which they can be used.

Examples:
Who's *looking after* (taking care of) the children?
He *looks down on* (disdains) all his relatives.
He *looked through* (examined) the accounts for her.
Things are *looking up* (improving).
He always *looked up to* (respected) her.

Look at the following list of some important phrasal verbs:

answer for	be responsible for
bank upon	depend on
be after	want
blow up	become angry
book in	reserve a place
break through	achieve something new
bring up	vomit
brush up	improve
call on	visit
cash in on	exploit
check in	arrive
check out	leave
chicken out	run away
come by	obtain
come down with	fall ill

come into	inherit
come up with	produce
cough up	provide money
cut in	interrupt
cut out	suitable
dig into	eat heavily
dish out	supply
fall in with	agree
fall under	come under
get at	obtain
get round	persuade
go along with	agree
go in for	like
have out	discuss
iron out	resolve
live up to	maintain
look down on	hold in contempt
look into	investigate
make out	understand
mess around	disturb
muddle along	manage somehow
open up	speak frankly
pass for	be taken for
play down	reduce
pull off	complete successfully
pull through	recover from illness, rough times
put in for	request
put up with	tolerate
rope in	persuade, include
run after	pursue
run down	disparage
run through	read briefly
see through	find the truth
stand for	represent
stand in	replace
stand up for	defend
take after	resemble

take to	begin to like
talk to	reprimand
touch upon	mention briefly

Idiomatic Expressions

Idioms are expressions peculiar to a language. They are fixed group of words the meaning of which cannot be inferred from an understanding of the component parts. They have only one form and can be used in only one way. English is very rich in idioms and the accurate and appropriate use of idioms is necessary for vividness and variety of writing and speaking.

Do not look for the literal meaning of an idiomatic expression. Thus, if somebody advices you not *to put all your eggs in one basket* or not *to count your chickens before they are hatched* he is not talking about poultry precautions but only cautioning you *not to risk everything on one attempt*, and also *not to depend on something before it materializes*. Thus, the special meaning attached to an idiom must be learnt carefully.

The form of an idiom is unchangeable and has to be learnt whole. It cannot be changed at will. Thus, the idiom *to rain cats and dogs* meaning to rain heavily cannot be changed to 'to rain dogs and cats' or 'to rain a cat and a dog'.

There are idiomatic combinations of a verb and adverb, or a verb and preposition (or verb with both adverb and preposition). Such combinations are called phrasal verbs as we have seen in the previous section. As in the case of idioms, their meanings cannot be guessed from the meaning of the verb alone. Apart from 'phrasal verbs', idioms are also formed by making use of other parts of speech. Here is a list of idiomatic phrases using the word 'mind' (noun):

1. *bear/keep* in mind: to remember
   Please bear in mind my words.
2. *be in one's right mind*: to be mentally well
   Are you in your right mind to suggest this?
3. *be of one mind*: to have the same opinion
   All the officers are of one mind about the problems.
4. *be of/in two minds*: to be undecided
   As the organizers were of/in two minds, it was not possible to come to a decision.
5. *be out of one's mind*: to be mentally ill
   The old man is out of his mind as seen by his behaviour.
6. *Call to mind*: to remember
   Try however hard, I am not able to call to mind the lines in question.

Important English Idioms

The following list presents a collection of important idioms employed most frequently in written and spoken English, today. The list contains only pure idioms i.e., fixed expressions whose meanings are quite different from the literal sense of the words. Semi-idioms, phrasal verbs and proverbs are not included in the list.

A to Z	thoroughly and completely
A B C of	the basic facts or principles of
Achilles' heel	a weak point in something that is otherwise without fault.
acid test	a decisive test of ability, etc.
add fuel to the flames	increase anger on any other strong active feeling.
all in all	on the whole, the main object of love and devotion
apple of one's eye	greatly loved person
ask for the moon	ask for something impossible
at loggerheads	quarrelling
at one's wit's end	very confused
at sea	in a state of ignorance or bewilderment
at the eleventh hour	at the last possible moment
back the wrong horse	make a wrong decision
be in the soup	be in trouble
beat about the bush	approach a matter in a round about way; speak indirectly
bed of roses	an easy, luxurious life
black list	list of persons or organizations regarded as untrustworthy
black sheep (of a family)	disreputable member
blow one's top	become very angry
blow one's own trumpet	praise one's own abilities
blue-eyed boy	favourite
bread and butter	living
by hook or by crook	by using any means possible
by leaps and bounds	very quickly
call a spade a spade	speak plainly and directly
carrot and the stick	reward and punishment
change one's tune	change one's opinion
climb on to the band wagon	join the victorious party
cock and bull story	an absurdly incredible story
close shave	danger just avoided
dead wood	something that is no longer useful
deliver the goods	achieve desired results
donkey's ears	very long time
drag one's feet	act very slowly

eat one's words	take back a statement
every Tom, Dick and Harry	everyone and any one
face the music	difficulties arising from something one has done
fish in troubled waters	try to gain an advantage from a confused state of affairs
fight tooth and nail	fight with great determination
free lance	person who acts independently
get cold feet	become nervous
get the boot	be dismissed
hat trick	triple success
hot water (be in, get into)	trouble
head and shoulders above	far superior
in black and white	in writing
in the long run	finally
jack of all trades	person who has the ability to do different kinds of work but none of it very well
keep one's fingers crossed	hope that nothing will go wrong
kick the bucket	die
last word	best example
live from hand to mouth	live from day to day
live wire	an active, lively person
look, stock and barrel	completely
maintain a low profile	be unassertive
lose heart	become discouraged
lose face	feel humiliated
much ado about nothing	make great fuss about a trifle
on one's toes	attentive
on the top of the world	happy and healthy
out of the blue	unexpectedly
out of the wood	out of danger
out of turn	out of the correct order
read between the lines	understand more than is actually written or spoken
red-tape	too many formalities/regulations
red herring	side issue to divert attention
stick out one's neck	take a risk
storm in a tea cup	a lot of fuss over a trivial affair
stone's throw	a short distance

take up the cudgels	support or fight strongly for something
throw up one's hands	give up hope
time of one's life	very enjoyable time
touch and go	a very risky situation
turn over a new leaf	make anew and better start
turn the tables	change a situation to one's advantage over one's enemy
up to one's ears	have too much work to do
wet blanket	one who by criticism discourages a person
wet behind the ears	lacking experience
weather the storm	survive a crisis
white elephant	an expensive and useless possession
with flying colours	very successfully
writing on the wall	warning of future.
wool gathering	day dreaming
worlds apart	very different, complete opposites

SINGLE WORD SUBSTITUTES

a place where animals are killed for food	abattoir
the repetition of the same sound at the beginning of two or more words	alliteration
a person who does something as a hobby and not for money	amateur
at home equally on land or in water	amphibious
the study of human society, customs, etc.	anthropology
a person who travels in space	astronaut
a person who does not believe in god	atheist
medical examination of a dead body	autopsy
a person with two wives	bigamist
a person with stubborn opinions	bigot
a person who leads an unconventional life	bohemian
a fussy government official	bureaucrat
a doctor who treats heart diseases	cardiologist
unthinking enthusiasm for one's country or cause	chauvinism
a person who sells sweets and pastimes	confectioner
a person with excellent taste and judgement	connoisseur
unlawful goods	contraband

a person who is recovering from illness	convalescent
a person who rouses people with emotional speeches	demagogue
a doctor who treats skin diseases	dermatologist
a person who has only slight interest in any subject	dilettante
a person who always thinks about himself	egoist
a doctor who treats diseases of glands	endocrinologist
a person who starts a commercial venture	entrepreneur
a person who enjoys eating	epicure
a person who is more outward than inward looking	extrovert
painless killing of people suffering from incurable and painful diseases	euthanasia
exact reproduction	facsimile
a person who sells flowers	florist
a list of words with meanings	glossary
a person who enjoys good food and drink	gourmet
a doctor who treats diseases of women	gynecologist
ranks of persona in order of importance	hierarchy
killing of one person by another	homicide
a person who constantly worries about his health	hypochondriac
a person who attacks popular beliefs	iconoclast
a person who comes into a foreign country to settle	immigrant
killing of children	infanticide
unable to pay debts	insolvent
dying without a will	intestate
a person who steals without intending to do so	kleptomaniac
a person who edits a dictionary	lexicographer
a person who gives his life for a cause	martyr
a person who hates women	misogynist
a fertile spot in a desert	oasis
a doctor who deliver babies	obstetrician
a doctor who treats eye diseases	ophthalmologist
a person who makes and sells spectacles	optician
a person who looks at the bright side of things	optimist
a doctor who treats diseases, and injuries of bones	orthopedist
a doctor who treats diseases of children	pediatrician
a doctor who studies diseases	pathologist
a person who always looks at the dark side of things	pessimist

a person who sells medicines	pharmacist
a person who collects postal stamps	philatelist
a person who poses to impress others	poseur
a doctor who treats mental diseases	psychiatrist
a story spread over a number of weeks	serial
not planned beforehand	spontaneous
a person who does not drink	teetotaler
act of quiet walking	stroll
any construction to commemorate great events or persons	monument
amount deducted from the declared price	discount
a man who accompanies another in a crime	accomplice
a building where dead bodies are kept before burial	mortuary
custom of having more than one wife at the same time	polygamy
collection and discussion of essays by several persons on a topic	symposium
cruel killing of a large number of people	massacre
custom of marrying only within the clan	endogamy
disease accompanied by pain , stiffness and inflammation of muscles and joints	rheumatism
easy solution for all problems	panacea
excessive formalities in official or public business causing unnecessary delay	red-tapism
expressing sympathy with people in their loss of their dear ones	condolence
group of small islands in a sea	archipelago
imitate others to cause amusement	mimicry
information of death given in newspapers with a brief history	obituary
institution where persons suffering from tuberculosis are given treatment	sanitarium
knowledge of everything	omniscience
of the same kind or nature	homogeneous.

Linkers or Thought Connectives

Read the following sentences and note the underlined phrase in them:

1. He engages himself in social work *in addition to* being the secretary of a flat owners' association.
2. She worked very hard to get that job, *however*, she was not lucky enough to get it.
3. The temperature appears to be stable, *although* it has been raining for the last two days.

4. He is good at mathematics, *but* weak in grammar.
5. She is good at painting *and* playing music.

The words/expressions underlined in the above sentences are called **linkers or thought connectives**. Using the correct linkers helps us in sequencing and linking up ideas and concepts in a piece of writing. Here are five sets of some **important linkers** that we use in our different forms of writing and speaking:

1. in addition to; further; moreover; apart from; although; however; though; in spite of;
2. whereas; on the contrary; for example; for instance; thus; such as ; in addition;
3. furthermore; then; in this case; indeed; surely; above all; certainly; in the same way;
4. on the other hand; in contrast; whereas; instead; similarly; more importantly;
5. additionally; in the same way, because, especially, then, of course, fortunately, before, after, besides, well, in other words, even; but.

Appendix B

Common Errors in English Communication

WRONG USE OF PREPOSITIONS/PREPOSITION PHRASES.

Unnecessary Use of Prepositions

There is a general tendency in most people to add prepositions after verbs where their use is unnecessary. In the following sentences, the underlined prepositions are unnecessary. You must try to avoid them in your speech and writing:

1. The manager admired for our home guard's bravery.
2. The security admitted him in.
3. They hope you would answer to their request soon.
4. My cousin approached to me for help.
5. The boss asked to the clerk why she was late.
6. The leaders attacked upon the Chairman's views.
7. The guard was awarded with a cash prize for his alertness.
8. The company could not bear up the burden of heavy taxes.
9. Computers have greatly benefited to the communication network in the country.
10. The management refused to bow down to the worker's demands.

Omission of Prepositions

Another tendency in many people is to omit prepositions in places where they are necessary. The following are some samples. They should be used with the preposition as given in brackets.

1. That is an idea we fully agree. (agree with)
2. She came and asked the car. (asked for)
3. The manager assured us loan. (assured us of)

4. We will have to bear her short temper. (bear with)
5. Would you care some juice? (care for)
6. We could not convey him the bad news. (convey to him)
7. They have to explain us the problem. (explain to)
8. If they go there, they will have to put up many difficulties. (put up with)
9. My colleagues presented him an expensive gift. (presented him with)
10. They will write us again soon. (write to us)

Use of Wrong Prepositions

The following sentences are examples of the use of wrong prepositions. The correct use is given in the brackets. Replace the preposition underlined with the one given in brackets against each sentence.

1. She was trying to aim her remarks against her friend. (at)
2. My boss is angry against me. (with)
3. That child is bad *in* spelling. (at)
4. That manager has great confidence about his secretary. (in)
5. She is very good in tennis. (at)
6. That thief confessed about his crime. (to)
7. That staff member has been charged of misappropriation. (with)
8. The boy is suffering with malaria. (from)
9. We want to dispense of his services. (with)
10. Will you sign the document with ink? (in)

USE OF WRONG TENSES

In the following sentences, the verbs are in the wrong tense. They ought to be replaced by the form of the verbs given in brackets against each sentence.

1. I am hearing that you are leaving for the West Indies next month. (hear)
2. He is forgetting to do his duties very often. (forgets)
3. We think to hold the tournament next month. (are thinking of holding)
4. They have booked our tickets yesterday. (booked)
5. He is playing cricket from his 6th class. (has been playing)
6. It happened when I stood in the bus stand. (was standing)
7. Ask him about the document when he will arrive. (arrives)
8. On Sunday next week, I am eating biryani. (am going to eat)
9. I am understanding the lecture with some effort. (understand)
10. she is remembering me very well. (remembers)
11. she is loving her friend. (loves)
12. I am having a headache. (have)

USE OF WRONG WORDS

One of the mistakes committed by many people is the use of inappropriate words in sentences because of the confusing nature of similar looking words or near synonyms. In the following sentences, the underlined words are to be replaced by the words in brackets.

1. She refuses to <u>hear</u> me when I give my suggestions. (listen to)
2. I liked the <u>travel</u> a lot. (journey)
3. The staff strike didn't have any <u>affect</u> on the boss. (effect)
4. He doesn't seem to <u>care for</u> his vehicle. (take care of)
5. <u>No one</u> of the letters has reached in time. (none)
6. This medicine seems to be very <u>efficient</u>. (effective)
7. She didn't <u>do</u> this mistake. (make)
8. He wanted to know the <u>last</u> score. (latest)
9. The boss doesn't seem to have any <u>principals</u>. (principles)
10. We shall discuss this issue <u>sometimes</u> tomorrow. (sometime)
11. They <u>shall</u> do all the arrangements. (make)
12. The parties reached a <u>wide</u> agreement on the issue. (broad)
13. He <u>denied</u> to accept the gift. (refused)
14. The patient was <u>released</u> last evening. (discharge)
15. This soldier is working on the <u>boundary</u> of the country. (border, frontier)

USE OF WRONG AGREEMENT

1. He <u>don't</u> know the problems involved. (doesn't)
2. They <u>doesn't</u> believe in superstitions. (don't)
3. I <u>doesn't</u> appreciate this technology. (don't)
4. You <u>doesn't</u> understand what I am going through. (don't)
5. They always <u>does</u> like that. (do)
6. I didn't <u>came</u> to know about the accident.(come)
7. They didn't <u>knew</u> about the function. (know)
8. She didn't <u>went</u> to the market yesterday. (go)

He She It	does doesn't
I You We They	do don't

He She It	works
I You We They	work

In these sentences, the singularity or plurality of subjects determines whether the verb will be singular or plural. In the sentences 9–13, we also find that when two subjects are joined by "and', the verb can be singular or plural, depending on whether the subjects are notionally singular or plural.

9. Dosa and chutney <u>are</u> what we generally have for breakfast. (is)
10. Sheela and Saheen <u>is</u> going shopping today.(are)
11. These days, a few hundreds rupees <u>are</u> a meager salary. (is)
12. The United States <u>have</u> the largest share of gold reserves.(has)
13. The audience <u>have</u> utmost respect for the performer. (has)

If the subjects in the sentence are joined by "either ….or", "neither…nor", the verb agrees with the subject nearest to it.(rule of proximity)

14. Either my sisters or my brother <u>are</u> coming. (is)
15. Neither the principal nor the students <u>is</u> interested. (are)

When two subjects are joined by connectors like "with", "along with", "as well as" etc., the first subject assumes more importance. Hence, controls the verb form.

16. The books, along with the stationery, <u>has</u> to be returned.(have)
17. The cow as well as the shepherds <u>are</u> not to be seen since yesterday.(is)

When the subjects are preceded by words like "each", "every". "one", "every-one", etc, the verb is always singular.

18. Every book and journal on the racks <u>are</u> torn. (is)
19. Each of the students in the college <u>want</u> to have a copy of the book.(wants)
20. One of the ships <u>have</u> a red mast.(has)
21. Everyone in this room <u>have</u> a valid reason to be angry. (has)

Appendix C

Practice Exercises

FILL IN THE BLANKS WITH THE RIGHT FORM OF THE VERB(S) GIVEN IN BRACKETS.

I Simple Present Tense

1. I ____________ cricket everyday. (play/plays)
2. My friend Raj ____________ movies always. (watch /watches)
3. The sun ____________ in the east. (rise/rises)
4. They ____________ the problem now. (understand/understands)
5. Hari always ____________ in the evening. (swim/swims)
6. Millie ____________ a cold. (have/has)

II Present Continuous Tense

1. I ____________ in the garden now. (play)
2. Sssh! Don't talk, someone ____________ about in the backyard. (move)
3. He ____________ now. (jump)
4. See there, all the boys ____________ . (shout)
5. Why ____________ you ____________ so loudly? (talk)
6. Jaleel ____________ his dinner now. (have)

III Present Perfect Tense

Wife : ____________ you ____________ all the rooms, Jaspal? (lock)

Husband : Yes, I have. And I ____________ all the doors and windows. (check) ____________ you ____________ the milkman to stop the milk? (tell)

Wife : Yes, I have. Where is the cat? ____________ you ____________ it to the man next door? (give)

Husband : No, I ____________ but I am sure it will be all right. (have not)

Wife : ____________ you ____________ the newspapers? (stop)

Husband : Yes, I have. ____________ you ____________ the thermos flask with coffee? (fill)

Wife : Of course, I have. ____________ we ____________ anything? (forget)

Husband : No, I am sure we haven't. Come on. Let us make a move. (The husband tries to start the car but it won't go.)

Wife : ____________ you ____________ to fill petrol? (forget)

Husband : My God! Yes, I have.

IV Present Perfect Continuous Tense

1. It ____________ for two hours. (rain)
2. I ____________ for you here for half an hour. (wait)
3. She____________ for a rich man all her life. (look)
4. Ever since Sheela refused to go to university, her father____________ her to get married. (press)

V Simple Past Tense

1. I ____________ the B.A. degree examination in 1968. (pass)
2. Mr Raheem ____________ in Hyderabad in 1948. (settle)
3. He ____________ a piece of land in the suburbs two years ago. (buy)
4. The Municipal Engineer ____________ for a bribe last week. (ask)
5. Jayram ____________ to Spain last year. (go)

VI Past Continuous Tense

This morning, I ____________ (walk) along the road. I ____________ (go) to the bus-stop to receive my friend. When I reached the bus-stop my friend had already arrived. He ____________ (wait) for me. We greeted each other and after a while, we started homeward. When we reached our house, my parents ____________ (wait) for us. Few minutes later we heard a cry from outside. All of us came out to know what the matter was. When we came out, some people ____________ (stand) in the middle of the road because of some accident.

VII Past Perfect Tense

1. I ____________ already ____________ my ticket, so I went into the theatre. (buy)
2. When the devotee ____________ in the river, he went back to the temple. (bathe)

3. The salesman showed her a new gadget yesterday, but she ____________ a new one the day before. (buy)
4. He reached the airport at seven, but the plane ____________ half an hour earlier. (land)
5. Her husband ____________ coffee before Reena got up. (prepare)
6. I ____________ him before he left for London. (see)

VIII Past Perfect Continuous Tense

1. When Mr Krishna came to the school in '85, Mr Rao ____________ already ____________ there for 5 years. (teach)
2. At 8 o'clock last night, I ____________ a book for two hours. (read)
3. When I saw him, he ____________ for his friend over half an hour. (wait)
4. Jeevan ____________ in New York for ten years before he moved to London. (live)
5. For a quarter of a year in 2009, he ____________ karate, the martial art. (practise)
6. Before India became independent, the British ____________ (rule) us for over 200 years.
7. Sandhya ____________ (teach) in a school for two years before she started teaching in an undergraduate college.
8. Venu ____________ (work) in a different organization for 12 years before he joined this company.

IX Simple Future Mood

1. I ____________ him tomorrow. (see)
2. Tomorrow ____________ Sunday. (be)
3. He ____________ me the money next week. (give)
4. If you ask him, he ____________ you. (help)
5. The machine ____________ if you switch it on. (function)
6. You ____________ a wonderful job if you work hard. (get)

Write suitable verbs forms choosing from the given alternatives

1. I have ____________ (came/come) from Africa.
2. We ____________ (doesn't/do not) know how to cook rice.
3. When we went to her house yesterday she ____________ (is/was) speaking to someone over telephone.
4. He ____________ (has/had) not finished the work that I gave him.
5. They have been ____________ (work/working) for the national company for the last five years.
6. She has ____________ (take /took/taken) leave for ten days.

Fill in the Blanks with the Correct Tense Form of the Verb Given in Brackets

1. Sharks __________ (live) in water.
2. The President __________ (speak) to the victims yesterday.
3. Have you ever __________ (be) to Puttaparthi?
4. I __________ (has) not completed my work yet.
5. He __________ (get) up daily at 5 a.m.
6. We __________ (drive) up to work generally.
7. The teams __________ (visit) the universities every five years.
8. The thieves __________ (leave) by the time the police arrived.

Read the Following Sentences and Fill in the Blanks with the Correct Form of the Verbs Given in Brackets:

1. Please be quiet, I __________ (try) to fix this machine.
2. Let's not go out now. It __________ (rain).
3. Excuse me, I __________ (look) for a phone booth. Is there one here?
4. Listen! Can you see those members __________ (discuss) the issue?
5. The financial situation is already bad and it __________ (worse) now.
6. The lab __________ (open) at 9 a.m. and __________ (close) at 6 p.m.
7. What time do the banks __________ (close) in Hyderabad?
8. The number of people without jobs __________ (increase).
9. I don't understand the word 'mantle'. What __________ it __________ (means)?
10. How often __________ the car __________ (break down)?

Complete the Following Sentences Choosing the Appropriate Word/Phrase from Among the Choices Given:

1. The Prime Minister __________ himself to help expand solar energy projects.
   (a) vowed (b) supported (c) pledged (d) swore
2. Proposals are being __________ for public comment.
   (a) laid (b) up (c) advised (d) drawn up
3. As soon as the monsoon __________ transport in the inner city becomes more and more difficult.
   (a) have started (b) had started (c) started (d) starts
4. I don't mind __________ the cook but I am not going to have anything to do with washing clothes.
   (a) to help (b) helping (c) help (d) for helping

5. Did you watch the last episode of the serial 'Hum' on the TV last Saturday? ________
   (a) No, I didn't. (b) Yes, I didn't. (c) No, I did. (d) I did.
6. All of a sudden, fire ____________ in my neighbor's farm.
   (a) broke away (b) broke through (c) broke down (d) broke out

Read the following sentences and fill in the blanks with an appropriate phrasal verb choosing from the list given below:

1. He ____________ the manager's proposal.
2. The statute ____________ only a single interpretation.
3. Why didn't you ____________ my warning?
4. One of the members ____________ of the agreement.
5. To ____________ such a contention would set a dangerous precedent.

(backed out; backed up; accede to; act on; admitted of; give up; cash in)

Look at the following phrasal verbs. Discuss their meanings with your group members and write individual sentences of your own using each of these phrases:

put aside; reach out to; gone through it; transform into; hold it down; peel off; turn up; hang about; mount on; tuck in; fit into; intrude on.

Read the following passage on Henry Ford and fill in the blanks with the correct verb forms:

Ford ______ (be) among the most successful motor companies in the world. Today its cars ______ (range) from the practical and the sporty to the luxurious. But it is for mass production and for ______ (make) the car affordable to everyone that its founder, Henry Ford, is best remembered. Ford was ______ (born) in 1863 on a farm in Michigan. He ______ (be) the eldest of six children. For all the farming families, life ______ (be) a struggle. Henry's father, William Ford was more successful than many Michigan farmers. Growing up on a remote farm, Henry soon ______ (show) signs that he belonged to a new generation of Americans ______ (interest) more in the industrial future than in the agricultural past. Henry ______ (hate) farm work and ______ (do) everything he could to avoid it. He was not lazy but his interest ______ (lie) in mechanics. He would set to work eagerly if he were ______ (give) a mechanical job. When he was twelve, he ______ (become) almost obsessively ______ (interest) in clocks and watches. Soon, he was ______ (repair) them for friends.

Read the following passage on 'help' and fill in the blanks with the correct verb forms:

It ______ (be) a terrible thing not to help others in need. However, such people who do not ______ (help) others are not many, but the few can ______ (be) very cruel. It is ______ (believe) by many that service to man is service to God and it ______ (seem) to be an important belief. These days, we ______ (find) many young people very forthcoming to help others. Here we are ______ (remind) of Vivekananda. His exact words ______ (be), "Unselfishness is more paying than selfishness, only, people do not ______ (have) the patience to practice it."

Read the following passage on the Stephensons and fill in the blanks with the correct verb forms:

George Stephenson and his son Robert Stephenson ______ (make) a tremendous contribution to the development of railways. George did not ______ (invent) the steam locomotive, but he ______ (ensure) that the insights and discoveries, which were ______ (begin) to come together from a variety of sources, were ______ (engineer) into a practical machine. In this context, a man of relatively humble origins ______ (become) one of the world's greatest engineers. He ______ (be) brilliantly pragmatic in ______ (ensure) that his son, Robert, was ______ (give) all encouragement in assisting the engineering of the early railways. The passage of time has ______ (do) nothing to pale the remarkable achievements of the two Stephensons (for it is unwise to separate father and son). Their inventive genius was most brilliantly ______ (display), probably during the year 1829–30, but this ______ (be) surely not the end. They ______ (stamp) their final mark on the shape of the locomotive when, in 1833, they ______ (add) a pair of carrying wheels behind the firebox to produce the first 2-2-2 engine 'Patentee'. This ______ (become) a benchmark passenger type in both France and Great Britain and was ______ (build) in the Great Britain until 1894. Their final major contribution ______ (be) in 1842, when they became the first to provide engines with link motion valve gear. This ______ (be) a fascinating invention, which was so extremely suitable for locomotives that it continued to be ______ (use) for new construction in the last phase of the great days of steam.

APPENDIX D

Model Question Papers

Model Question Paper – 1

Max. Marks 100

I. Choose ONE of the topics given below and participate in a group discussion. (25 marks)

1. Computer is only a tool. Its intelligence is a myth.
2. Students should be tested on only the subjects of their choice at the end of the B.Tech course.
3. Teacher attitude in the classroom is more important then the knowledge he/she imparts.
4. Spiritual quotient is necessary for success.
5. Evolution is not about progress.

(Note to teachers: a GD can be conducted on one of the topics to test the students' spoken skills in groups of 6–7 students. The evaluation for this discussion could be based on the following parameters;

- *initiative, leadership* – 10 marks
- *language, fluency, logicality, connectivity* – 10 marks
- *team work* – 5 marks.)

II. What are the important elements in any human communication? Explain with the help of examples. Keep the difference between the spoken and the written mode in mind while answering this question. (Marks – 10)

III. Given below are a number of situations. State which of the elements of human communication is predominant in each of them. Explain why. (Marks – 10)

1. A telephonic conversation.
2. A teacher talking to students, trying to explain why it is important for them to use the right kind of study skills.
3. A session where you have to say how you found a certain course useful, interesting.
4. A group discussion where you have to express your views on whether government funds should be diverted for primary or higher education.
5. A conversation with your principal, where you are trying to convince him about not detaining students with less attendance.

IV. Read the paragraph given below and prepare notes. Maintain alignment between the main and the sub points. (Marks –15)

Abbreviation is a shortened form of a word or a group of words. An abbreviation is used in writing to save time and space. Some abbreviations are also used in speaking. Abbreviations are often made by taking the first letter of the word or the first letter of each important word in the group and writing it in capitals. For example, PO stands for post office and COD for collect on delivery or cash on delivery. Sometimes, the first letter is printed as a small letter followed by full stop, as in the case of m. for married and b. for born. In some cases, other letters of the word are also added. Thus, ms. means manuscript and ft. means foot. Sometimes abbreviations are made from the initial letters or syllables of a group of words and they spell out a word, such as NATO or WAVES. Such abbreviations called acronyms are not followed by a full stop. Letters used in abbreviations are sometimes doubled for the plural form, as in the case of ll., used for lines, and pp., used for pages.

For certain abbreviations, small capital letters may be used instead of large capitals. The abbreviations AD, BC, AM, and PM are often printed in small capitals. Of course, abbreviations are most often used to stand for common words such as the names of days, months, of states, and of countries. Long words and phrases are often abbreviated too, such as Lieut., for lieutenant and RFD for rural free delivery. Titles and academic degrees are usually abbreviated; DD is used instead of the longer Doctor of Divinity, and HRH instead of His (or Her) Royal Highness. In modern business Co. means Company; Ltd. means limited.

V. Give the structure of a feasibility report. State only the outline of the structure. (Marks – 5)

VI. Answer any three of the following short answer question topics. (Marks – 3 × 5 = 15)

1. Write a short note on the need for creativity in communication.
2. Distance and positioning as body language in a communicative situation.
3. Routine reports.
4. Speed reading.
5. The don’ts in a telephonic conversation.

VII. Answer the following questions relating to grammar

1. Look at the following sentences and identify the main and the auxiliary verbs in each. (Marks – 5)

a. The opinions of the people were not revealed.
b. The glaciers in the North and the South Pole are melting fast.
c. The air is getting polluted very fast.
d. The rare birds had left their habitat before the team of experts could reach them.
e. I have broken my hand.

2. Fill in the blanks by using the right form of the verb given in brackets. (Marks – 5)

a. The shopping mall _______ (open) at 9am everyday.
b. We often _______ (see) English movies.

c. I have a jeep but I ______ (do) drive it often.
d. The river Amazon flows into the Pacific Ocean.
e. How often ______ (does) you write to your parents?

3. A friend of yours has just come back from a holiday. You are asking her about it. Make complete sentences, using the clues given, as answers given by your friend (Marks – 5)

a. stay/cousin ______________________
b. place/see/many ______________________
c. meet/old friends ______________________
d. visit / famous temple ______________________
e. walk up / the mountain ______________________

4. In the sentences given below, fill in the blanks by putting the verbs in the correct form – Present perfect or the Present perfect continuous. (Marks – 5)

a. Look! Somebody ______ (spoil) the freshly painted wall.
b. I smell cooked food. Have you ______ (cook).
c. Shekhar is an actor. He ______ (appear) in several films.
d. I ______ (read) the book you gave me. But I ______ (finish) it yet.
e. I ______ (lose) my key. Can you help me look for it?

Model Question Paper – 2

Max. Marks 100

I. Make a presentation for about 15 minutes on any one of the given topics. (Marks – 25)

1. Creating a sustainable environment.
2. Rainwater harvesting.
3. Waste management.
4. Modern lifestyle and growing pollution.
5. The relevance of engineering science for rural India.

(Note to the teacher: This can be a PowerPoint presentation where students are encouraged to take material from various sources and do a little bit of research on the topics. Marks can be divided for:
- *Language and delivery* – 10 marks
- *Organization and presentation of material* – 10 marks
- *Relevance of material* – 5 marks.)

II. What are the different types of interpersonal communication? Explain with examples. (Marks – 10)

III. Given below are expressions that are related to negotiation situations. Explain what paradoxes they reveal, and. how. (Marks – 2 × 5 = 10)

1. 'How boring. I wish I could do this work soon. I don't mind doing it all myself.'
2. 'This was my idea. But now, the entire team will get credit for it.'

IV. You work for a reputed bakery. There is a proposal to open a branch of the bakery in another part of the city. You have been sent to study the feasibility of such an extension. Given below are some of the conditions you have observed. You think the venture will be profitable. Write a formal report conveying the same to your proprietor. (Marks – 20)

- People in this part of the city are already aware of the popularity of the bakery.
- It has quite a few colleges around. The young mass will be the regular clients.
- The area is also surrounded by residential complexes with largely an upper middle-class population.
- The number of working women in the locality seems to be fairly large.
- There is a good access road to the place and a shopping complex with pre- existing parking facility.

V. Write answers to any three of the following question topics. (Marks– 3 x 5 = 15)

1. Johari window and interpersonal communication.
2. Types of listening.
3. Basics about conducting and participating in a meeting.
4. Factors that block creativity.
5. Hearing and listening.

VI. Answer the following questions relating to grammar

1. Make sentences using the words in brackets. Use them in the Simple past or the Past continuous form. (Marks – 5)

a. (phone/ring/have/a shower)
My phone ________________________
b. (watch/a film/television/news flashed)
We were ________________________
c. (sleep/lightening/struck)
I was ________________________
d. (discuss/ the issue/ director/walk in)
We were ________________________
e. (listen / music / guests / arrive)
I was ________________________

2. Fill in the blanks, using verbs of your choice in the appropriate passive form. (Marks – 5)

a. No decision on this matter can __________ till next morning.
b. The book will have to __________ as soon as possible.
c. The injured man __________ to the hospital.
d. It is customary for the luggage to __________ by the customs officials.
e. We all wanted to __________ before day break tomorrow.

3. Fill in the blanks, using either the simple present or the present continuous (be –ing form) of the verbs put in brackets. (Marks – 5)

a. Arun is in India now. He __________ (stay) at the Taj Krishna. He generally __________ (stay) there whenever he visits.
b. My parents __________ (live) in Delhi now. But we are from Hyderabad.
c. I __________ (teach) English. But this semester, I __________ (teach) Psychology too!
d. I generally __________ (drive) the bike. But because of backache I __________ (use) the car these days.
e. He never __________ (drink) alcohol at home.

4. Fill in the blanks using the right form of the verbs given below. (Marks – 5)

a. Jim was not at home when I reached.
He __________ (already / leave)
b. The man was a complete stranger to me.
I __________ (never/ see/before)
c. I was very tired when I reached home.
I __________ (play / tennis / wholemorning)
d. Mr Mathur no longer has his car.
He __________ (sell / last month)
e. I invited Margaret for dinner. But she could not come.
She __________ (promise / someone else)

Model Question Paper – 3

Max. Marks 100

I. Read the following situation and negotiate appropriately. (Marks – 25)

This is a situation where a group of engineers with different specializations are meeting the government agencies for the finalization of a low cost housing project. The group constitutes young bright engineers who are very enthusiastic to take up this work and carry out certain experiments. But the government has certain objections. It is also possible that they would like to give the project to a well-known agency they have worked with before.

(Teacher's note: Organize a negotiation situation around this theme. You could conduct the negotiation session with students as the group, which wants to take up the project and the staff members as the representatives of the government who are apparently examining the proposal. Please take the help of the staff teaching engineering subjects for this. You could ask each of the students to face the negotiation situation individually or in groups of two or three. Mark them for:

- *The negotiation strategy they are adopting.*
- *The manner in which they are manipulating the language, tone, intonation, etc.)*

II. What is mind-mapping? Explain how it can be used productively for essay/project writing and note taking in the classrooms. (Marks – 20)

III. What are the features of a well-written business letter? You are the manager of a financing company that finances business ventures. A group of fresh B.Tech graduates have taken your help to set up a computer firm. Even a month after they should have paid the first installment, you have heard nothing from them. They have been reminded once by your company. Write a firm second reminder. Complete your letter with names of company, etc. (Marks – 20)

IV. Answer any three of the following question topics. (Marks – 3 × 5 = 15)

1. Minutes writing.
2. Active listening.
3. Positives (Do's) in a telephonic conversation.
4. Language of technical writing.
5. Importance of eye movement and chunking while reading.

V. Answer the following questions relating to grammar

1. Change these sentences into reported speech, changing the words where necessary. (Marks – 5)

a. "I'm listening to the programme", he said.
b. "I stayed awake the whole night because of headache", she said.
c. "We can phone home once we reach the station", they said.
d. "I have to see you tomorrow", my manager said.
e. "I met her about three months ago", the engineer told.

2. Complete the following passage by filling in the correct tense form of the verbs given in brackets. (Marks – 6)

There ______ (is) a lot of excitement before the songs competition ______ (begin) yesterday. All the participants ______ (are) busy ______ (do) last minute rehearsals. When it ______ (start), there was silence everywhere. One by one, the participants ______ (stand) on the dais, faced the audience and ______ (sing). When my turn ______ (come) I ______ (overcome) my nervousness by trying to focus only on my rendering. I ______ (do) a fairly good job. My joy ______ (know) no bounds when the results were ______ (declare). Looking back, I now feel happy that I won it.

3. Given below are some sentences. Read them and write down another sentence with the same meaning. Begin the sentence the way it has been suggested and note whether you are using the active or the passive voice. (Marks – 4)

a. All the clients may write their suggestions in the register kept at the counter.
Suggestions ______________________________

b. The Principal postponed the meeting due to his ill health.
The meeting ______________________________

c. Somebody might have issued the book if it is not here.
The book ______________________________

d. A short circuit might have caused the breakdown.
The breakdown ______________________________ .

IV. Read the sentences given carefully and make new sentences with similar meanings. Use the reported speech. (Marks – 5)

a. "Listen to the music with eyes closed", he told us all.
b. "Don't you ever come on Sundays", he told him strictly.
c. "Open the door please", my brother told me.
d. "My parents are arriving tomorrow", she said.
e. "We visited her this morning", the nurse told.

Model Question Paper – 4

Max. Marks 100

I. Answer the following questions in about 150 words each. (Marks – 5 × 5 = 25)

1. What are the different factors one has to be sensitive about to become an effective speaker?
2. During a presentation, how does a speaker ensure audience involvement?
3. What preparations are necessary before giving a presentation?
4. What are the factors crucial to an effective group discussion?
5. What is the importance of turn-taking during a group discussion?

II. Answer any four of the following questions in about 150 words each. (Marks – 4 × 5 = 20)

1. What is Information Technology?
2. What are the effects of MIS on organizations?
3. What is communication?
4. What factors are necessary for effective communication?
5. What is creativity in the context of day-to-day communication?
6. What is the function of language?
7. What are the elements of human communication?

III. Given below are some situations in which we do reading. State what kind of reading you would do in ANY 5 of these situations? Remember, depending on your need you would read the same material differently at different points of time. So answer this, given your present condition and necessities. (Marks – 5 × 4 = 20)

1. The TV guide for the week.
2. An English grammar book.
3. An article in a popular magazine about the famous actors of yesteryears.
4. The weather report in your local newspaper.
5. A bus timetable.

6. A fax that has come for you at the college office.
7. A recipe.
8. A short story by your favourite author.

IV. Answer any three of the following question topics in about 100 words each.

(Marks – 3 × 5 = 15)

1. Radiant thinking.
2. Process description in technical writing.
3. Basics of minutes writing.
4. 'Johari window'
5. Different levels of communication

V. Answer the following questions relating to grammar

1. Fill in the blanks using the right form of the verbs given below. (Marks – 5)

a. Rajesh was not at home when I reached.
He __________ (already / leave)

b. The person was a complete stranger to me.
I __________ (never/ see/ before)

c. I was very tired when I reached home.
I __________ (play / cricket / whole morning)

d. Mr Mukesh no longer had his car.
He __________ (lost / last month)

e. I invited Farhana for dinner. But she could not come.
She __________ (promise / someone else)

2. In these sentences you have to talk about your future plans. Use the right words to answer these questions. (Marks – 4)

a. i. Which car are you going to buy?
ii. I am not sure. I __________ (Maruti Suzuki)

b. i. Is Raghu coming with us to the picnic tomorrow?
ii. He said he would try to. He __________ (even/bring/his sister) along with him.

c. The computer is troubling me a lot. I __________ (take it/ workshop.)

d. i. Where are you going to be transferred to?
ii. I __________ (have to/shift/Bangalore).

3. Fill in the blanks, using verbs of your choice in the appropriate passive form.

(Marks – 5)

a. No decision on this matter can __________ till next morning.
b. The book will have to __________ as soon as possible.
c. The injured man __________ to the hospital.
d. It is customary for the luggage to __________ by the customs officials.

e. We all wanted to __________ before day break tomorrow.

4. Match the following phrasal verbs with their correct meanings from the list given below in brackets and make sentences with any six of the phrases.

(Marks – 6 × 1 = 6)

a. get round
b. go along with
c. go in for
d. have out
e. iron out
f. live up to
g. look down on
h. look into

(stop, agree, overcome, discuss, hold in contempt, dislike, investigate, like, maintain, persuade, resolve)

Model Question Paper – 5

Max. Marks 100

I. Answer the following questions in about 150 words each.

(Marks – 5 × 5 = 25)

1. What are the elements of human communication?
2. What is empathy?
3. Each of us is naturally endowed with the capacity to read and analyze body language. Discuss.
4. What is mirror imaging? State its role in the communication situation.
5. What are the implications of the Johari window?

II. Answer any four of the following questions in about 150 words each.

(Marks – 4 × 5 = 20)

1. In a communication situation, what is the ratio of verbal vis-à-vis non-verbal communication? State in short the various factors involved in communication.
2. How is the nature of a specific interaction determined?
3. What are the different kinds of basic emotions the face can express?
4. Can we always interpret cross-legged gestures to be defensive gestures? Give reasons for your answer.
5. How many significations does the 'arms behind the back' posture have? State any two with examples.
6. Define the concept of negotiation.

III. Answer any FIVE of the following questions in about 100 words each.

(Marks – 5 × 5 = 25)

1. How can we define listening?
2. What is the difference between hearing and listening?

3. Give a brief account of the different kinds of reading.
4. Describe the SQ3R technique as an active reading process.
5. What are the features of a well-written business letter?
6. What does 'Pointing" mean? What can you infer from a person's feet pointed towards the door?
7. What is the difference between "scratching the neck" and "stroking the neck"?
8. Mention some important factors that determine the success of a negotiation.

V. Read the following text and fill in the blanks with the appropriate words from the list given below the text. (Marks – 10)

Ambiguity is an inevitable and ________ (1) part of our lives and thinking. The state of ambiguity is natural for most of us most of the time. But this becomes ________ (2) for the computer because it can function only when these ambiguities are ________ (3) and it is given very ________ (4) instructions as to what is to be done. But should issues always be defined clearly? And even if it can be done, don't they often ______ (5) their critical edge? Considering the ________ (6) perspective, issues have to be unambiguously ________ (7) to make the computer work but it also means that we are forced to ________ (8) to the computer when we use it. From the view of the man-machine ________ (9), the victims of inequality here are the human beings. This problem cannot be ________ (10) unless the computer is developed enough to be able to handle ambiguity.

(Practical, interface, indispensable, critical, precise, conform, defined, solved, resolved, lose, danger, façade, manage, outrun)

VI. Read each sentence to find out whether there is any error in any underlined part. (Marks – 10)

1. She insisted on (a) my agreeing (b) to meet her in the evening (c) to discuss about (d) the new topic. No error (e).

2. Who could deny (a) that he had a remarkably (b) kindly (c) disposition that won him (d) friends wherever he went ? No error (e)

3. One of the conditions laid down (a) are related to (b) the extra expenditure he keeps (c) talking about (d). No error (e).

4. At college (a), I had (b) many choices but I did not hesitate (c) to offer (d) Mathematics and Music. No error (e).

5. Giving the examination at such a short notice will at best get me a third. No error.
   a b c d e
6. It was disgraceful that someone should attempt to wreck a project aimed for
   a b c
   giving relief to the unemployed. No error.
   d e
7. There can be little doubt that the problem of unemployment is growing; the
   a b
   statistics speak in themselves. No error.
   c d e
8. In some areas children moving from primary to secondary schools are still
   a
   elected according to their current level of academic attainment. No error.
   b c d e
9. The traditional Western attitude towards work makes full-time employment
   a
   and the desire for promotion the central motivating force in life. No error.
   b c d e
10. It is by now increasingly recognized that workers may be thrown out by
    a b c
    industrial forces beyond their control. No error.
    d e

VII. Read the following text and fill in the blanks with the appropriate words from the list given. (Marks – 10)

I like everything about my motorbike, ________ its color and speed. ________ recently, its pickup has decreased. I don't know why. I'll take it to a mechanic ________ (it becomes worse). Most mechanics these days are undependable, but ________ my mechanic is reliable ________ (being economical). Actually, ________ I bought my bike in 1988, I have been having a good driving time, except ________ for an occasional problem here and there. It has been giving me a decent mileage. I feel the mileage will increase further in a few days time, ________ I'm taking that extra little care of it ________, I'll have to wait and see. But in the meantime, a friend of mine advised me to sell the bike and buy a different model. I don't think I'll do that, at least not in the near future. I will use it for five more years and ________ I'll sell it, may be. But the idea is very disagreeable to me.

(List of words: but, because, especially, then, of course, fortunately, before, after, besides, well, in other words, even, always, beforehand, afterwards, give up)

Bibliography

Andrea J. Rutherford, *Basic Communication Skills for Technology*, Pearson Education, Inc. New Delhi, 2001.

Anne Dannellon, *Team Talk: The Power of Language in Team Dynamics*, Harvard Business School Press, Boston, Massachusetts, 1996.

Antony Jay and Ros Jay, *Effective Presentation: How to be a Top Class Presenter*, Universities Press (India) Limited, 1999.

Ashraf M. Rizvi, *Effective Technical Communication*, Tata McGraw-Hill Publishing Company Ltd., New Delhi, 2005.

Daniel Goldman, *Emotional Intelligence*, New York, Bantam Books, 1995.

Friedrike Klippel, *Keep Talking*, Cambridge University Press, London, 1984.

Georgi Lozanov, *Outline of Suggestology and Suggestopedia*, London, Gordon and Breach, 1978.

G. V. Raj, *Vocabulary Expansion and Reading Comprehension*, Gamaliel Publishers, Hyderabad (unpublished, limited edition for private circulation)

Hari Mohan Prasad and Rajnish Mohan, *How to prepare for Group Discussion and Interview*, 2nd Edition, Tata McGraw-Hill Publishing Company Ltd., New Delhi, 2005.

Jean M. Hiltrop and Sheila Udall, *The Essence of Negotiation*, Prentice-Hall of India, New Delhi, 1997.

Jeremy Harmer and John Arnold, *Advanced Speaking Skills*, Essex, Longman Group Limited, 1978.

Joseph O'Connor and John Seymour. *Introducing NLP: Psychological Skills for Understanding and Influencing People*, Thorsons, Hammersmith, London, 1990.

K. K. Sinha, *Business Communication*, Galgotia Publishing Company GPC, New Delhi, 1999.

Lewis Hedwig, *Body Language: A Guide for Professionals*, Response Books (a division of Sage Publications India, Pvt. Ltd.,) New Delhi., 1998.

Matthew M, Monipally, *Business Communication Strategies*, New Delhi, Tata McGraw-Hill Publication Company Limited, 2001.

N Krishnaswami, *Modern English: A Book of Grammar, Usage and Composition*, McMillan India Limited, 1974.

Patricia Hayes Andrews and Richard T. Herschel, *Organizational Communication: Empowerment in a Technological Society*, Houghton Mifflin Company, USA, 1998.

R. C. Sharma and Krishna Mohan, *Business Correspondence and Report writing*, Tata McGraw-Hill Publishing Company Ltd., New Delhi, 1978.

R. P. Bhatnagar and Bhargava Rajul, *English for Competitive Examinations*, Mcmillan India Limited, 1989.

Richard Palmer, *Brain train*, London, Spon Press, London, 2003.

Rob Nolasco and Lois Arthur, *Conversation*, Oxford University Press, London, 1987.

Robert Heller, *Essential Managers: Communicate Clearly*, Dorling Kindersley, London, 1998.

Saroj P. Karnik, *Comprehensive Business Communication*, Orient Longman, 1977.

Satyanand Saraswati, *Yoga Nidra,* Monghyr, Bihar School of Yoga, 1976.

Stephen R. Covey, *7 Habits of Highly Effective People*, Pocket Books, 1999.

Steven E. Pauley, Daniel G. Riordon, *Technical Report Writing Today,* Houghton Mifflin Company, USA, 1999. (Published in India by All India Traveler Bookseller, Delhi, 1999)

Terry Burrows, *Creating Presentations*, Essential Computers, Dorling Kindersley Books, 2000.

The Secrets of Body Language, (A book produced by the Readers Digest Association Limited), London. 1977.

Tim Hinhle, *Negotiating Skills*, Essential Managers, Dorling Kindersley Books, 1998.

Tony Buzan, *The Mind Map Book,* BBC Consumer Publishing, 1993.

V. R. Narayanswamy, *Strengthen your Writing*, Orient Longman, 1979.

William C. Miller, *Flash of Brilliance: Inspiring Creativity Where You Work*, Reading, Massachusetts, 1999.

Wycoff J., *Mindmapping, Your Personal Guide to Exploring Creativity and Problem Solving*, Berkley Books, New York, 1991.